Critical Reflections on Regional Competitiveness

Since the early 1990s, governments and development agencies have become increasingly preoccupied with the pursuit of regional competitiveness. However, there is considerable confusion around what exactly regional competitiveness means, how it might be achieved, whether and how it can be measured, and whether it is a meaningful and appropriate goal for regional economies. The central aim of this book is to provide a comprehensive and critical account of these debates with reference to theory, policy and practice, and thus to explore the meaning and value of the concept of regional competitiveness.

The book is structured into three parts. Part I introduces the concept of regional competitiveness by tracing its origins and exploring its different meanings in regional economic development. This will critically engage with political economy approaches to understanding the nature and dominance of the competitiveness discourse. Part II interrogates the pursuit of regional competitiveness in policy and practice. This critically evaluates the degree to which the pursuit of competitiveness is encouraging convergence in policy agendas in regions through an examination of key determinants of policy sameness and difference, notably benchmarking and devolved governance. Part III explores the limitations to regional competitiveness and explores whether and how its predominance in the policy discourse might be challenged by alternative agendas such as sustainable development and wellbeing. This focuses on the developing qualitative character of regional development.

This volume critically engages with the theory and policy of regional competitiveness, thus providing the first integrated critique of the concept for undergraduate and postgraduate students, as well as academics interested in regional development and policy. It will unpack the concept of regional competitiveness and explain its usefulness, limitations and policy appeal, as well as examining its sustainability in the light of evolving governance structures and the imperatives of broadening regional development agendas.

Gillian Bristow is a Reader in Economic Geography in the School of City and Regional Planning, Cardiff University. Her research interests focus on regional economic development, governance and policy. This book develops her seminal critique of regional competitiveness published in the *Journal of Economic Geography* in 2005.

Routledge Studies in Human Geography

This series provides a forum for innovative, vibrant, and critical debate within Human Geography. Titles will reflect the wealth of research which is taking place in this diverse and ever-expanding field.

Contributions will be drawn from the main sub-disciplines and from innovative areas of work which have no particular sub-disciplinary allegiances.

Published:

Critical Reflections on Regional Competitiveness

Theory, policy and practice

Gillian Bristow

Routledge
Taylor & Francis Group

LONDON AND NEW YORK

First published 2010
by Routledge
2 Park Square, Milton Park, Abingdon, Oxfordshire OX14 4RN

Simultaneously published in the USA and Canada
by Routledge
711 Third Avenue, New York, NY 10017, USA

First issued in paperback 2014

Routledge is an imprint of the Taylor & Francis Group, an informa business

Typeset in Times New Roman by Taylor & Francis Books

British Library Cataloguing in Publication Data
A catalogue record for this book is available from the British Library

Library of Congress Cataloging in Publication Data
Bristow, Gillian.
 Critical reflections on regional competitiveness : theory, policy and
practice / Gillian Bristow.
   p. cm. – (Routledge studies in human geography ; 31)
 Includes bibliographical references and index.
 1. Regional economics. 2. Competition. 3. Regional planning. I. Title.
 HT388.B75 2009
 338.900917–dc22
                    2009021463

ISBN13 978-1-138-86732-1 (pbk)
ISBN13 978-0-415-47159-6 (hbk)

Contents

PART III
Moving beyond competitiveness   119

Illustrations

Preface

I have been thinking about competitiveness for a long time. Competitiveness is clearly something everyone instinctively and intuitively understands – we're all engaged in competition of various kinds and for various things – and we can understand its logic when applied to various different kinds of activity, whether in the business world or in a sporting context, in particular. But competitiveness is now everywhere and has become an enormously elastic concept which has been stretched to apply to nations, regions and cities – to places more generally. It is certainly deeply embedded in the things I am interested in around how regions work, their development, their governance structures and policy approaches. Competitiveness dominates the thinking around regions, it is central to the key metrics of regional development and performance, and it provides a seemingly unalterable focus for regional strategies and policy. And yet there are huge questions around what 'regional competitiveness' actually means, whether in fact regions and places can be conceived as 'competitive' in the same way that firms or football teams can, and what the enduring desire to be competitive means for the policies shaping regions in practice. It is the desire to articulate these questions and to attempt to find some answers to them that has prompted the writing of this book, and which has ultimately shaped the direction it has taken.

Acknowledgements

It would be impossible to thank everyone who has influenced my ideas and helped me with the writing of this book, but I would like to acknowledge a few. Staff and students in the School of City and Regional Planning, Cardiff University have provided a constant source of new ideas and thoughts. An initial period of study leave was also invaluable in providing me with the thinking space and reading time needed to start this project, whilst numerous discussions in seminars, lectures and coffee breaks with MSc Regeneration Studies students, colleagues in the Urban and Regional Governance research group, as well as with PhD students and other staff have proved to be hugely influential in very many ways in shaping its development. I am also hugely appreciative of all the support received from the first-class administrative and technical staff in the school – thank you for bearing with me during the busiest writing times and providing invaluable practical support at particular crunch times (office removals included). Special thanks must go to Kevin Morgan for constant encouragement, inspirational advice and mentoring, and numerous pointers to relevant sources and readings. I have learned a huge amount from you, thank you. Also thanks for Thomas Berger for assistance with Chapter 4 and the deconstruction of competitiveness metrics and league tables. More widely, I have also benefited from seminar discussions at a number of universities, including Lancaster and Liverpool. I am also very grateful to all at Routledge, especially Michael P. Jones and Andrew Mould for helping me see this project through from beginning to end.

And last but by no means least, I continue to be inspired, supported and encouraged by a fantastic family (the Bristows, Aldworths and Kitchens) to whom I owe an enormous debt of thanks – Phil, Al, Rich, Dave, Paula, Gareth, Rhi, Carys, Ieuan, Gethin, Lowri and Joe – you are all very special people. To Mum for always being there, and for all the conversations, love and laughter we share – thank you. This is for you, and in memory of Dad – our rock and inspiration and always remembered. And finally, Mart – you're an absolute star, my best mate and I can't thank you enough for your constant good-humoured and laid-back encouragement and support throughout this. You've been amazing. Thanks for all the cheering up conversations when out walking with Rafa, and for sharing all the wonderful distractions provided by the ups and downs of supporting the Bluebirds and Liverpool – you mean the world. YNWA.

Abbreviations

BCI	Business Competitiveness Indicators
BERR	Department for Business, Enterprise and Regulatory Reform (UK)
BISW	Bundesländer im Standortwettbewerb (Benchmarking German States Index)
CAG	Competitiveness Advisory Group
CBI	Confederation of British Industry
CPE	cultural political economy
DTI	Department of Trade and Industry (UK)
EDA	Economic Development Administration (US)
ERT	European Roundtable of Industrialists
ESPRIT	European Strategic Programme for Research and Development on Information Technologies
EU	European Union
GDP	gross domestic product
GVA	gross value added
ICTs	information and communications technologies
IMD	International Institute for Management Development
IMF	International Monetary Fund
ISEW	Index of Sustainable Economic Welfare
MNE	multinational enterprise
NGO	non-governmental organization
NUTS	Nomenclature of Territorial Units for Statistics
OECD	Organisation for Economic Co-operation and Development
OMC	Open Method of Co-ordination
RDA	Regional Development Agency
SRA	strategic relational approach
UKCI	United Kingdom Competitiveness Index
UNCTAD	United National Conference on Trade and Development
UNICE	Union of Industrial and Employers' Confederations of Europe
WAG	Welsh Assembly Government
WAVE	Wales a Vibrant Economy
WDA	Welsh Development Agency
WEF	World Economic Forum

Part I

The discourse of regional competitiveness

Introduction
Neoliberalism and the regional competitiveness hegemony

Introduction: the competitiveness hegemony

The dynamics of economic, social, political and cultural change in the contemporary world are increasingly shaped by the pursuit and promotion of global competitiveness. Indeed, competitiveness has become a 'hegemonic discourse' within public policy circles in developed countries (Schoenberger, 1998). International organizations ranging from the International Monetary Fund (IMF), the World Bank and the Organisation for Economic Co-operation and Development (OECD) are all busy urging governments everywhere to reform the business climate, promote investment and stimulate competitiveness. Furthermore, the pursuit of competitiveness has been elevated to primary strategic importance in the Lisbon strategy of the European Union (EU) which explicitly states its aim as being 'to make the EU the most competitive and dynamic knowledge-based economy in the world capable of sustainable economic growth with more and better jobs and greater social cohesion' (CEC, 2000: p. 2).

This preoccupation with competitiveness is premised on certain pervasive beliefs, most notably that globalization has drastically changed the structural properties of the global economy and that best practice governance is secured through neoliberalism. Neoliberalism can be defined as 'a distinctive political-economic philosophy that took meaningful shape for the first time during the 1970s, dedicated to the extension of the market (and market-like) forms of governance, rule and control across – tendentially at least – all spheres of social life' (Peck and Tickell, 2007: p. 28). Neoliberalism's assertion that economic policies favouring supply-side innovation, competitiveness, decentralization, deregulation, privatization and the promotion of the active, 'workfare' state make for good governance has become widely accepted across all scales of governance and across all parts of the world, albeit in slightly different forms. It has, in short, become the hegemonic contemporary form of liberal society (Leitner et al., 2007).

Neoliberalism has entailed a 'reshuffling of the hierarchy of spaces' associated with Fordist-Keynesian national forms of regulation and, concomitantly, to the mobilization of new institutional arenas such as regions and cities which are deemed to be the 'breeding grounds' for the development of new productive forces (Lipietz, 1994: p. 36). Indeed, by the 1990s the 'region' quickly emerged as a determinate 'space of competitiveness' (Brenner, 2000), meaning it had widely been identified as a key territorial zone and institutional arena for the promotion and pursuit of competitiveness strategies. The pursuit of regional competitiveness as a policy goal has been adopted with particular enthusiasm by the EU and by national governments across the developed world. It has risen to particular prominence in the UK, where the pursuit of regional competitiveness has moved to centre stage in the policy statements of national government, and where Regional Development Agencies (RDAs) have been explicitly tasked with the responsibility for making their regions 'more competitive' and akin to benchmark competitive places such as Silicon Valley (HM Treasury, 2001). Similarly in Germany, the debate around regional competitiveness has produced what Hickel (1998) refers to as a kind of 'locational hysteria' in which regional political and economic actors have become obsessed with the structural competitiveness of their jurisdictions relative to other European and global locations. In short, the competitiveness hegemony is such that, according to certain analysts, 'the critical issue for regional economic development practitioners to grasp is that the creation of competitive advantage is the most important activity they can pursue' (Barclays, 2002: p. 10).

Policy documents extolling the importance of competitiveness tend to present it as an entirely unproblematic term and, moreover, as an unambiguously beneficial attribute of an economy. This is particularly the case at the regional scale. Competitiveness is portrayed as the means by which regional economies are externally validated in an era of globalization, such that there can be no principled objection to policies and strategies deemed to be competitiveness enhancing, whatever their indirect consequences. For example, the European Commission (CEC, 2004: p. viii) states that 'strengthening regional competitiveness throughout the Union and helping people fulfil their capabilities will boost the growth potential of the EU economy as a whole to the common benefit of all'. Similarly, the UK government sees its regional policy as being one of 'widening the circle of winners in all regions and communities' (DTI, 2001: p. 4).

The region as a space of competitiveness?

Notwithstanding this policy enthusiasm, there is considerable confusion as to what the concept of regional competitiveness actually means and how, if at all, competitiveness can be applied to regions in a way that helps in understanding their economic performance in a manner akin to firms. Indeed, in a manner cognate with debates around clusters (see Martin and Sunley, 2003), policy acceptance of the existence and importance of regional competitiveness appears to have run ahead of a number of fundamental theoretical and empirical questions.

Academics have recently begun to ask some key questions around the subject, with a number of publications providing an important foundation for enhanced debate (e.g. Bristow, 2005; Martin et al., 2006). However, to date these contributions have essentially focused on understanding the regional competitiveness concept, its meaning and basis in pertinent theoretical (ostensibly economic geography) debates. This has done much to highlight the inherent slipperiness of the competitiveness concept but has paid only limited attention to broader questions such as why the pursuit and benchmarking of competitiveness has become so salient at the regional scale, whether and to what extent regions have the collective unity and capacity to develop coherent competitiveness strategies or to challenge the competitiveness hegemony, and how sustainable the region is as a space of competitiveness, given the emergence of newer sub-national state spaces such as city-regions. Yet these are questions of growing importance, given the obdurate persistence of economic inequalities between regions, the dynamic pace and change of state restructuring and rescaling processes, not to mention highly pressing concerns about the social and ecological limits of narrowly economistic approaches to development.

In short, the central objective of this book is to critically interrogate and unpack the region as a 'space of competitiveness' in its constituent economic and political terms. In other words, the aim is to critically reflect on whether and how the region is 'competitive' in terms of being an appropriate space both for moulding vibrant economic activities, and for instituting effective structures of governance and forming coherent and effective competitiveness strategies. In so doing, it will seek to provide new insights into regional studies and competitiveness debates and highlight the contested and contingent nature of the competitiveness discourse in practice, as well as examining the scope for the region itself to mould, resist and perhaps ultimately challenge the competitiveness hegemony.

The need to consider regional competitiveness in broader political economy terms is justified by the evolving nature of regional studies, and in particular in the emergent, more fluid understanding of the meaning and importance of the 'region' coupled with the ongoing reproduction and rescaling of the competitiveness orthodoxy.

Understanding the 'region'

The widespread belief in the concept of regional competitiveness carries the implicit assumption that 'the region' is both clearly understood and unequivocally defined. This is manifestly not the case and regional geographers have long struggled to define the boundaries of their fundamental object of study, such that what actually constitutes a region remains an object of mystery (Harrison, 2006) and a vibrant source of ongoing debate.

In exploring the enduring ambivalence around 'the region' in geography, Paasi (2002) makes an analytical distinction between three ideas of the region that geographers perpetually lean back on: pre-scientific, discipline-centred and critical approaches. The pre-scientific view implies that the region is a practical choice, a given spatial unit (statistical area, municipality or locality) which suits the purpose of collecting or representing data, but which has no particular conceptual role. This view is typical of applied and comparative research. The discipline-centred view of regions sees them as objects or results of the research process. As such they are often formal or functional classifications of empirical elements and used to legitimate a particular geographical perspective. More recently, geographers have favoured a more critical approach which asserts that regions are not natural or self-evident entities. Instead, they are fundamentally social constructions or putative 'imagined communities' based on ideational and identitive qualities. As Jayasuriya (1994: p. 412) observes, 'regionalism is a set of cognitive practices shaped by language and political discourse, which through the creation of concepts, metaphors, analogies, determine how the region is defined; these serve to define the actors who are included (and excluded) within the region and thereby enable the emergence of a regional entity and identity'. Since there are multiple identitive qualities or ways of seeing a region, there are many different approaches to defining its boundaries. Regions may be defined in relation to their economic processes, their commonalities of production patterns, labour markets and innovation networks. Alternatively, regions may be defined in relation to their political processes, their administrative and territorial functionality and sense of identity. Or regions may be defined with respect to their cultural processes, their homogeneity of norms, traits, associations and practices.

Within the literature on regional competitiveness and innovation, the term 'region' has been typically understood as a given scale between the national state and the local. This may clearly pertain to very different geographical entities and indeed, within competitiveness debates the 'regional' label has been variously applied to sub-national territories and geographical areas as diverse as the US states and the Canadian provinces (which are often larger than many European countries), as well as to small-scale industrial districts and areas such as NUTS II regions in Europe that do not necessarily correspond to any single jurisdiction (Doloreux and Parto, 2005). Nevertheless, within these debates the 'region' has developed a close association with governance and territoriality and with the economic processes deemed to function most effectively at the sub-national scale.

The traditional idea of regions as purely bounded spaces has been increasingly challenged by the relational perspective which sees regions, in whole or in part, as open, discontinuous 'spaces of flows' constituted by an extended network of social relations and thus fundamentally shaped by their relations and connectivity with other scales and sites of economic organization (see Allen et al., 1998; Hudson, 2007; Lagendijk, 2007). This approach sees regions as open, porous and unbounded spaces, defined by their linkages and relations within networks and not predefined by any particular territorial boundary. It also asserts that regions are comprised of individual actors (e.g.

workers, managers, consumers and politicians), as well as collective ones (such as firms, governmental bodies and other organizations). These actors are embedded in various structures of socio-institutional relations and actor networks which influence their decisions and actions and lead them to pursue multiple (economic and non-economic) goals and strategies. As such, the relational perspective also emphasizes the interdependencies between institutions, individuals and wider social structures, and the mutual constitution of the local and the global. There is thus no such thing as an encapsulated local. Economic agents travel around the world and develop international business networks. The local labour market is shaped by international migration, and the local knowledge base is informed by daily international news reports and information received from around the world. Global forces clearly exist and develop inside the local (Bathelt, 2006; Massey, 2007a).

This has important implications for policy. A relational reading of regions implies that policies for regions need to be attuned to their international flows and connectivities and not simply to their static regional characteristics. It also implies that regions are not entirely autonomous entities capable of determining their own futures, although there is room for local initiative. Instead, it asserts that regional policy has to go beyond a narrow territorial basis and enter wider policy arenas to negotiate arguments and claims. It also suggests the need to encourage local activation as a process of enrolment in, rather than protection against, the global (Bathelt, 2006).

There is, however, a powerful argument suggesting that it may be more productive not to view relational and territorial approaches as binary opposites or competing either/or choices, but rather to view them from a both/and perspective shaped by theoretical, methodological and political context (Hudson, 2007). In particular, Jones and MacLeod (2004) have drawn attention to the distinction between 'regional spaces' and 'spaces of regionalism' – the former referring to the regionalization of economic activity, and the latter to processes of political mobilization around cultural expression and increased civic identity. Thus, they argue that the relational perspective is more convincing when applied to economic flows and interchanges, but is less pertinent when understanding spaces of political regionalism, where political action is mobilized territorially (see also MacLeod and Jones, 2007; Painter, 2008). In short, actually existing regions are likely to be mutually constitutive of both territorial and networking processes.

New regionalism

The reification of the regional scale as a space of competitiveness is inextricably linked to the emergence of the New Regionalism school of thought which, during the course of the 1990s, became something of an orthodoxy in policy-related regional development studies (see Lovering, 1999). Indeed, as Webb and Collis (2000: p. 858) explain, 'the starting point for New Regionalists is almost always the concept of regional competitiveness'. This school of thought embodies two related sets of ideas (for an overview see Harrison, 2006). The first component of New Regionalism provides theoretical insights from political sciences and is concerned with the emergence of the regional state and sub-national forms of governance as a viable and forthright partner in the search for a successful compromise between state and market (Keating, 1998). Keating argues that Western Europe has experienced a progressive new regionalism since the mid 1980s whereby the authority of the nation-state has been increasingly challenged by the twin forces of internationalization from above, and the assertion of regions and civil society from below. In this regard, the nation-state is being 'hollowed out' and its role as the prime regulator and controller of economic governance increasingly eroded (Jessop, 1994).

Whilst state governance is acknowledged to be a major context for region (and identity) building, international markets and political responses to global capitalism increasingly provide the impetus for regionalization and the accentuated importance of regions (Paasi, 2002). Thus, the second and dominant thesis of New Regionalism is that of economic new regionalism. This focuses on the historico-empirical claim that in the context of globalization and the emergence of new spaces of competition (Brenner, 2000), such as continental Europe, 'the region' has become the crucible of economic development and wealth creation. Using insights gained from endogenous growth theory, institutional economics and cognitive psychology, proponents of economic new regionalism assert that the critical propellants of competitive advantage lie increasingly in facets of the regional business environment, which acts as the breeding ground for the development of new production forces (Lipietz, 1994) and as the container within which firm-level innovation is created and sustained (Porter, 1990; 2003).

The resurgence of regional economies at a time when the forces of globalization were regarded by many as having reduced the world to a placeless mass was examined in seminal work by Michael Storper (1997). Storper powerfully argued that regions (or more accurately, core regions) are critical to economic interaction, innovation and wealth creation in the global economy through their ability to foster traded (input–output) and untraded (institutional and social) interdependencies between firms. Where these interdependencies are localized, the region becomes the 'key, necessary element in the supply architecture for learning and innovation processes' (Storper, 1997: p. 22). Furthermore, success in knowledge creation and innovative learning processes can lead to an economic territorialization whereby an activity's economic viability is rooted in assets that cannot easily be replicated or imitated in other places. In so doing, Storper played a prominent role in promoting the rediscovery by many academics of the regional scale and in propelling a research agenda focused on the role of specific regional rules, conventions, norms and trust in stimulating the innovation capacity critical to economic success (e.g. Morgan, 1997; Malmberg and Maskell, 2002). This was strengthened and supported by the acknowledgement that the most successful and dynamic regional economies in North America and Western

Europe at this time were characterized by a set of specific institutional and social conditions.

New Regionalism quickly emerged as a fashionable banner for reifying the region as an object of study and creating a powerful but essentially normative narrative that the region had somehow attained a new authority in and of itself (Lovering, 1999). Its appeal lay in its seeming ability both to offer a convincing theoretical explanation for the success of fast-growth regions such as Silicon Valley, Emilia-Romagna and Baden Württemburg and to establish the region as a new, effective arena for locating the institutions of post-Fordist political-economic governance (Harrison, 2006). Yet this very appeal also provided the focus for much of its subsequent and most vehement criticism (see, for example, Lovering, 1999; MacLeod, 2001; MacKinnon et al., 2002).

These critiques focus on three salient issues. The first of these is the inherent tendency within New Regionalism to read off institutional developments from successful regional spaces and assert their generic relevance. For Lovering, New Regionalism is a highly selective amalgam of 'all things good' in the world's regional economies, such that as a consequence it is simply 'a set of stories about how parts of a regional economy might work, placed next to a set of a policy ideals which might just be useful in some cases' (Lovering, 1999: p. 384, original emphasis). Whilst Harrison (2006) usefully demonstrates that this indictment may be doing New Regionalism something of a disservice, there remain concerns about the failure of new regionalists to accurately represent their theoretical abstractions and their innate subtleties to the policy-making community.

The second key criticism is New Regionalism's tendency to focus narrowly on the processes of developing soft institutionalism and business growth *within* a region, thereby ignoring the constitutive outside influences, social networks and institutional relations impacting *upon* regions and shaping their development trajectories. As indicated above, the relational perspective sees patterns of regional development and prosperity as reflecting relations of power and control over space, where core regions tend to occupy dominant positions and peripheral regions play marginal roles within wider structures of accumulation and regulation. In this regard, each regional economy is in a distinct position, since each is a unique mix of relations over which there is some power and control and other relations within which the region may be in a place of subordination (Massey, 2004). This perspective posits that certain regions, such as the South East of England in the UK, are likely to develop a hegemonic political and economic position which not only shapes their own development but also impacts upon the development processes of other regions.

The third critique of New Regionalism is that it represents 'a poor framework through which to grasp the real connections between the regionalization of business and governance and the changing role of the state' (Lovering, 1999: p. 391). In short, it fails to engage with the national state and, more often than not, with the local state also. Recent research has begun to highlight how the rise of the regional scale is deeply intertwined with the restructuring of the state, and there has been growing emphasis on understanding regional devolution, in particular, as part of a broader process of state reorganization (see, for example, Goodwin et al., 2005). Yet, as Harrison (2006) observes, expansion on these ideas remains necessary, with a need to better understand the relationship between the region and the nation-state and the complex way the national state produces, reproduces and articulates the scalar and spatial sites of economic governance. In particular, he argues for deeper interrogation of the way in which the nation-state is becoming involved in meta-governance or the government of governance through the processes of democratic devolution, constitutional change and functional decentralization to regional institutions such as development agencies. This raises questions as to whether this results in the nation-state building new capacity and steering the regions more effectively, or whether instead it retreats from support of the region and does less. It also brings into sharp focus the need to understand whether regions and the people within them are merely subjects of policy and economic development processes, or objects of policy endowed with the agency to shape, develop and deliver policy (Hudson, 2007).

The multi-scalar nature of competitiveness

The ongoing debate around the role and significance of regions is running in parallel with continuing processes of reproduction and rescaling of the competitiveness orthodoxy. Competitiveness has not only come to dominate supranational, national and regional policy agendas, but also increasingly frames debates around local (particularly urban) economic development (see, for example, Buck et al., 2005; Hooper and Punter, 2006). It has a particularly prominent place in the revitalized city-region agenda taking hold across the UK and much of Europe which sees cities and their wider hinterlands as a new space of competitiveness and a critical territorial platform for contemporary capitalist success.

The emergence of the city-region raises new questions about the role of regions and whether and how city-regions relate to regions and share similar processes of construction and evolution (Harrison, 2008a). Certain parallels are indeed clear. Like regions, the growth of city-regionalism can be understood as a deliberate process of scalar reorganization and crisis management by the state, reflective of its ongoing search for an appropriate scale on which to perform key economic growth functions in the midst of capitalism's inherent tendency towards uneven development. However, it may also in part reflect local and regional interests and the demand for greater autonomy provoked by this unevenness in development and the inequities that result (Harrison, 2007). This suggests the construction of city-regions, like regions, is inextricably linked to the reconfiguration of the state, which is in turn rendered more complex and perhaps contested as a result.

Notwithstanding these questions, the successful rescaling of the competitiveness orthodoxy reveals much about the slippery nature of the competitiveness concept, as well as its enduring power and appeal. The debate as to what the definitive 'space of competitiveness' is or, indeed, if there even is one, is far from resolved. Michael Porter, a key advocate of the competitiveness concept, has played a powerful role in asserting the importance of the region to competitiveness. However, in his work he also implies that the concept is generally applicable to all places or 'locations', including cities, regions and entire nations (see, for example Porter, 1995; 1998; 2003). Others tend to identify particular scales and spatial specificities as being more important than others in determining competitive success (see, for example, Begg, 2002; and Turok, 2004 for the importance of cities; and Camagni, 2002 for the importance of regions). This suggests there is no natural, pre-defined spatial scale at which issues of competitiveness are best theorized or analysed, with the sources of competitiveness originating from a variety of geographical scales, from the local, through regional, to national and even international (Cambridge Econometrics et al., 2002). However, the notion of competitiveness may take on a very different meaning depending upon the scale at which the term is being applied and the different microeconomic (firm-based) and macroeconomic (economy-wide) definitions of the term which then take precedence, thus creating an imperative for rigorous interrogation of how competitiveness is understood and pursued across different scales.

Competitiveness is thus a concept of enormous elasticity inasmuch as not only is what it means to be competitive a matter of debate, but also multiple entities, ranging from individual firms to entire economies, can be deemed to be competitive or not (Rosamund, 2002). What is particularly salient, however, is that whilst the meaning of competitiveness may vary across scales, the elevated status it occupies in economic development discourse remains a remarkable constant.

Critical questions

In summary, the competitiveness hegemony is such that it possesses the innate ability to be reproduced and rescaled such that it dominates development discourses across multiple spatial scales. This raises important questions about its specific definition and resonance at the regional scale, where it appears to have developed particular force and appeal. Yet regions themselves are in a state of flux, given the emergence of new sub-national forms which mean they are now one of several, plausible sub-national tiers within increasingly polycentric systems of governance. Regions are also being reconceptualized as broader social and material constructions, as products of both processes of territorial construction and their inherent relationships with other regions and scales, and thus as complex relational entities with an inherent variety and heterogeneity of interests. This nascent 'relational regionalism' (Harrison, 2008a) challenges the notion that regions will be necessarily coherent entities with a predetermined collective status and agency, and suggests that the prioritizing of the region as the pivotal scale and tendential unit for development and policy purposes should be based upon carefully grounded research rather than thin empirics (Hudson, 2007).

Thus, for a number of reasons the time has come to be reflective about the meaning, role and status of the region as a competitive space (or a container for economic activity and the sustenance of competitive firms), and a space of competitiveness (or a political object or subject of competitiveness strategies and policy). This is thus the purpose of this book.

The book is organized as follows. Part I focuses on problematizing the dominant policy discourse around regional competitiveness with reference to theory. It begins by examining what exactly is meant by the concept of regional competitiveness and explores the polysemous yet overlapping meanings of regional competitiveness used in academic debates and policy discourse (Chapter 1). Chapter 2 then examines how and why a discourse with ostensibly thin and ill-defined content has assumed such significance in policy circles. This critically engages with political economy approaches to understanding the nature and dominance of the competitiveness discourse, and its rhetorical power and usefulness for particular groups with particular interests. Part II interrogates the pursuit of regional competitiveness in policy and practice. This critically evaluates the degree to which the pursuit of competitiveness is encouraging convergence in policy agendas in regions through an examination of key determinants of policy sameness and difference, notably benchmarking and competitiveness metrics, and devolved governance. Part III explores the limitations of regional competitiveness for policy and practice and explores whether and how its predominance in the policy discourse might be challenged by alternative agendas such as sustainable development, resilience and well-being. This focuses on the developing qualitative character of regional development.

This volume critically engages with the theory and policy of regional competitiveness, thus providing the first integrated critique of the concept for undergraduate and postgraduate students, as well as academics interested in regional development and policy. It will unpack the concept of regional competitiveness and explain its usefulness, limitations and policy appeal, as well as examining its sustainability in the light of evolving governance structures and the imperatives of ever-widening regional development agendas.

1   What is regional competitiveness?

Introduction

National competitiveness has become a much-debated and much-maligned concept. Why some nations prosper and others do not has been one of the central questions in economics since the days of Adam Smith, yet there remains considerable debate as to whether nations can be deemed 'competitive' entities in the same manner as can firms. Certain authors highlight that countries do not go out of business in the same way as firms do and do not engage in trade as a zero-sum game. Rather, the success of one country creates opportunities for others (Cellini and Soci, 2002; Krugman, 1994). Others argue that all 'places' clearly engage in competitive behaviour in respect of the attraction of investment and key resources (Boschma, 2004; Turok, 2004). However, in spite of the growing literature on this topic there is precious little agreement as to what place competitiveness actually means or how it can be conceptualized.

The notion of regional competitiveness is particularly contentious and is characterized by a striking paradox that makes it a fascinating and rich focus for study. On the one hand, the region appears to have become a determinate space of competitiveness inasmuch as the rescaling processes associated with the spread of neoliberalism have enhanced its economic and institutional role within competitiveness agendas (see Introduction). It is indeed at the regional scale that many of the 'soft' factors that enhance the productivity of firms and raise general economic performance are deemed to be created and sustained.

Yet at the same time, the region represents a somewhat awkward scale for analysing what is essentially a concept concerning relative economic performance. This is because the regional or meso scale is neither macro-(national) nor microeconomic (firm-based). Regions are not like firms, inasmuch as they are not direct economic 'actors' with discrete control over the activities taking place within them. Neither are they like nations, since they do not possess the same macroeconomic policy levers. Nor do they go out of business. They may, however, be more vulnerable than nations to changing patterns of trade if they become over-specialized in particular economic activities which subsequently emerge elsewhere. Furthermore, regions do appear to compete for economic activities and use a range of material and immaterial inputs such as housing, infrastructure, communications and social networks to create or enhance their locational advantage which, in turn, may influence the competitive advantage of their firms (see, for example, Malecki, 2004). However, precisely how this is then transmitted into improved regional economic performance is not clear.

Regions are, in effect, the space in between the micro and the macro and are neither simple aggregations of firms nor scaled-down versions of nations (Cellini and Soci, 2002). The regional scale is thus possibly the most difficult and complex level at which to analyse competitiveness, not least because, as Budd and Hirmis (2004: p. 1021) observe, 'regional competitiveness appears squeezed between the rock of the national competitiveness debate and the hard place of the plethora of the volume of work on territorial competitiveness at an urban scale'. In short, the tendency to conflate *competitiveness* and *competition* at the regional level creates huge scope for analytical, conceptual and operational confusion. This raises interesting questions as to how and in what ways regions compete or are competitive, and whether either the firm-based or more macro export-orientated or income-based conceptions of competitiveness often used at the national level can be suitably applied to regions. The purpose of this chapter is thus to explore these questions and seek to find some way through the conceptual chaos which surrounds regional competitiveness.

Conceptions of regional competitiveness

Broadly speaking, there are two ways in which regional competitiveness can be understood – firstly, in the relatively narrow sense of competing over market share and resources, and secondly, in the much broader respect of the determinants and dynamics of a region's long-run prosperity. One of the major difficulties in understanding regional competitiveness is that these different conceptions typically tend to get muddled together and confused.

Regions in competition

Competitiveness is usually used to refer to relative firm performance. The discourse of firm competitiveness comes from two principal sources. The first of these is the discourse of the economics profession, where competitiveness is regarded as a somewhat abstract quality conferred upon successful firms by the markets within which they operate. Thus, 'the market is the impartial and ultimate arbiter of right behaviour in the economy and competitiveness simply describes the result of responding correctly to market signals' (Schoenberger, 1998: p. 3). The fusion of economics with evolutionary theory has imbued the concept with the notion of 'survival of the fittest'. Firms, like organisms, are seen as living on the edge, with survivors being those that are able to survive or 'win' in a dynamic world of economic competition

(Sheppard, 2000). Competitiveness has thus become inescapably associated with ideas of fitness and unfitness, and these in turn with the implied premise of merit, as in 'deserving to live' and 'deserving to die'. Second, competitiveness is also the discourse of the business community, where it represents the fundamental external validation of a firm's ability to survive, compete and grow in markets subject to international competition. This provides a pervasive and powerful means of explaining almost any behaviour, i.e. a firm must do 'X' in order to be competitive (Schoenberger, 1998).

Thus, at the level of the firm, competitiveness has a relatively clear meaning and refers to the capacity of the firm to compete, grow and be profitable in the marketplace. In principle at least, relative firm competitiveness can also be measured on a common scale. This is because it refers to commensurable units (firms) engaged in commensurable activities (competing in a market). Since it is perceived to reflect a firm's ability to survive competition and grow, firm competitiveness is generally conceived of in terms of output-related performance indicators such as productivity.

The globalization discourse has played a powerful role in the sedimentation of the idea that regions or places are equivalent to corporations, competing for market share within an increasingly interconnected and fiercely competitive global economy. Yet the literature on regions strongly disputes the notion that regions are engaged in some sort of global race in which there are only 'winners' or 'losers' (Kitson et al., 2004).

Clearly, whilst some regions, like firms, grow faster than others and enjoy changing relative shares of economic wealth and activity, regions are manifestly different from firms in a number of crucial respects. A major difference is that firms enter and exit markets whereas regions do not. In other words, the process of market selection drives certain firms out of business, whereas in regions, centralized public finance systems act to cushion the impact of economic decline. In other words, whereas firms face a distinct bottom line and may go out of business if they are uncompetitive, places do not.

Furthermore, unlike firms, the competitive success of one place is not necessarily at the expense of another. Indeed, regions and cities are often locked in complex interdependencies and networks of relations, some of which are cooperative rather than necessarily conflictual or competitive. They create markets for one another, people often commute between them, and supply chains typically cross their boundaries. This applies to cities too. As Unwin (2006: p. 5) observes, 'nobody would claim that if Manchester's economy performs poorly, this is good news for Liverpool. The demand for Liverpool's goods and services will shrink and Manchester will no longer be able to supply it with goods and services as cheaply or of the same quality.'

A further difference is that, unlike firms, regions may have more than one objective and are not driven simply by the pursuit of economic success or profit (Turok, 2004). Furthermore, they do not have direct control over all their assets and liabilities and so cannot respond as clearly to the changing incentives provided by the economic performance of their rivals. There may be signals encouraging them to limit growth, and there may be a diverse array of political interests and pressure groups to satisfy through their actions. In short, the effects of competition are moderated by a range of other resource allocation mechanisms.

Nevertheless, regions *do* engage in various kinds of emulous, competitive behaviour in a direct effort to maintain or enhance their economic position, and it is in this regard that the scope for confusing competition and competitiveness resides. Lever and Turok (1999: p. 792) observe that 'cities and other places compete with one another. This takes many different forms – some direct head-to-head competition for particular projects or events; others more indirect, subtle and incremental in nature' (see also Chien and Gordon, 2008). Indeed, the regional economic development literature is littered with examples of egregious strategic competitive behaviour and spatial 'contests' between regions. Regions, or more specifically their institutions and jurisdictions, compete in ever more sophisticated and complex ways for a number of economic inputs, including domestic and foreign direct investment (Tewdr-Jones and Phelps, 2000), highly skilled labour (Rohr-Zanker, 2001), shares of finite government resources (Morgan, 2001) and mobile capital and tourists, increasingly through mega-events such as the Olympics and Football World Cup (Jones, 2001). In addition, they engage in a range of activities designed to improve or enhance the locational assets which make it possible for them to attract and keep investment and migrants – that is, to become 'sticky places' (Markusen, 1996). This has spawned growing interest in the nature and type of assets deemed to be most significant, with contributions emphasizing factors such as bohemian diversity for attracting the 'creative classes' (Florida, 2002) to more generic 'quality of life' or socially embedded attributes (see Jessop and Sum, 2000; also Donald, 2001). By definition, these assets are difficult to analyse and measure and there is limited knowledge of how they emerge, develop and evolve over time. They are nevertheless given increasing importance in policy approaches to local and regional development.

The critical point here is that the conception of competitiveness as place attractiveness subtly changes the subjects of competition. This is very clearly articulated by Fougner (2006: p. 175), who observes, 'the primary subjects of the competition in relation to which the concept of international competitiveness is increasingly used have changed from firms to states. In accordance with this, the primary governmental problem on the part of state authorities is no longer to make firms more competitive, but to make the state itself more competitive ... rather than being complementary and interdependent, the two conceptions of international competitiveness in question belong to two different imaginary worlds.' If states are increasingly conceived as competitive entities, then it is easy to see how the places over which they have jurisdiction can be understood in similar terms.

This sort of competitive activity and the place marketing it engenders is particularly prominent at the city and city-regional scale and there is a burgeoning literature on the nature and consequences of this (see, for example,

Begg, 2002; Malecki, 2004; Buck et al., 2005). Indeed, in this regard there may be some justification for describing urban entities as competitive in a manner similar to firms, inasmuch as the combination of functional speciali- zation and the agglomeration benefits of urbanization means that cities and city-regions do compete in more discrete and similar markets and that gains to one city are more likely to be at the expense of another. In short, it is possible to generate a balance-sheet of city assets and liabilities. This is much more difficult at the regional scale, since regions are not engaged in such direct competition with one another and they are much less likely than cities to have functional specialization across their entirety, or to possess a unity of purpose between economic and social interests. Furthermore, regions also have to mediate the international division of labour, national political inter- ests and the diverse ways these factors are played out at the regional scale. Thus, whilst they may compete for a range of economic activities, the med- iation processes involved are typically very complex and often incomplete. Put simply, it is much more difficult to produce a balance-sheet of regional assets and liabilities (Budd and Hirmis, 2004).

Moreover, this sort of competition between regions may not necessarily enhance the performance of firms or improve regional economic performance or prosperity in the long term. Evidence of the significance of these locational assets as sources of improved firm performance and regional prosperity is thin at present (Buck et al., 2005) and this sort of competitive behaviour can itself have negative consequences (see Markusen, 2007; see also Part III of this volume). What is clear is that it is very easy to confuse and conflate *compe- tition* between regions for resources and assets with the *competitiveness* of those regions, and yet the two notions are subtly different. Whilst regions may compete with one another in certain ways, this is not to say they are compe- titive in the same way as firms are, where rivalry is more closely associated with clearly defined selection, incentive and thus performance outcomes. However, inasmuch as the competition between regions occurs for factors which help shape the development and sustenance of a business environment conducive to competitive firms, there are clear connections with the second, much broader conceptualization of regional competitiveness, which sees it as per- taining to regional economic prosperity. It is to this broader conceptualization that this chapter now turns.

Competitiveness as the dynamics and determinants of regional prosperity

An alternative, much broader perspective of regional competitiveness has come to prominence which conceptualizes it in terms of the determinants and dynamics of regional performance or prosperity. There is, however, consider- able academic disagreement as to whether competitiveness can be directly equated to a region's macroeconomic performance, or whether instead regio- nal competitiveness is 'revealed' through the microeconomic performance of the region's firms. What is also evident is that there is no one single, comprehensive theoretical framework underpinning this notion of competitiveness, but rather a range of different theoretical contributions which help to explain its particular dynamics and determinants.

For some authors, regional competitiveness is directly equivalent to regional macroeconomic performance. One of the principal proponents of this approach is Michael Storper, who defines regional competitiveness as:

> the capability of a region to attract and keep firms with stable or increasing market shares in an activity, whilst maintaining stable or increasing standards of living for those who participate in it.
>
> (Storper, 1997: p. 264)

This definition has quickly gained widespread academic acceptance and use, particularly amongst new regionalists (see Maskell and Malmberg, 1999; Malecki, 2002; Huggins, 2003). It derives from the discourse of firm-based competition and is in fact a direct application of a widely used definition of national competitiveness (see Chapter 2). The implicit assumption made, therefore, is that a region is a scaled-down version of a national, macroeconomy and that the notion of national competitiveness can be simply read off, rescaled and applied to a region. This is not an unreasonable assumption to make, given that regional economies are likely to be more open (to trade) than the national economies of which they are a part, and firm performance (through exports to international markets) has long been viewed as critical to regional prosperity (Kitson et al., 2004).

This conceptualization sees regional competitiveness as a combination of the competitiveness of a region's firms (defined in terms of their overall external validation through growth in market share) and a region's overall economic performance (or validation through sustained or improved levels of comparative prosperity). It therefore conceives of competitiveness in output-related terms and asserts that regional competitiveness and regional prosperity are in fact interdependent notions, if not directly equivalent. This assertion, and the definition's avoidance of equating regional competitiveness with productivity, is deliberate, as is explained by Huggins (2003: p. 89):

> Although low labour costs may initially contribute to the attraction of business investment to an area, such costs are in many ways a 'double-edged sword', resulting in employees working for lower wages than their counterparts in other localities and regions. Therefore, it can be argued that true local and regional competitiveness occurs only when sustainable growth is achieved at labour rates that enhance overall standards of living.

To put it another way, microeconomic productivity is considered to be a necessary but not sufficient condition for financial returns, increased market share or, ultimately, improved macroeconomic performance. A region may have a stock of highly competitive firms in the microeconomic sense, but if they are engaged in activities that create low added value per worker, then the economy will not be competitive in the macroeconomic sense. Thus, a region is 'competitive' according to this view when it has the ability to raise its standard of living and sustain 'winning' outcomes. This means it has the conditions to enable it to generate high profits and high wages.

The difficulty lies in understanding precisely what these 'conditions' are. At the national scale, there is considerable confusion around this. Krugman (1997a: p. 7) alludes to this confusion with his assertion that national competitiveness is typically defined rather vaguely as 'the combination of favourable trade performance and something else'. National competitiveness is thus variously conceptualized as reflecting a nation's ability to sell goods in the global marketplace, its ability to earn money – which refers to its overall 'results' in macroeconomic terms – its ability to adapt through innovation to changing market demands, and its ability to attract outside investments in financial capital and the skilled human resources required for development (see Berger, 2008). Thus, the 'something else' which Krugman refers to seems to relate to the uncertainty already alluded to surrounding the ways and means by which enhanced firm competitiveness shapes overall economic outcomes.

This confusion is amplified further at the regional scale. Clearly the factors that might operate to enhance profits and wages for national economies, such as exchange rate movements and price–wage flexibility may not exist or be as significant at the regional level. In contrast, other factors, such as the mobility of labour and capital between places, can impact hugely on regional conditions and performance (Camagni, 2002).

Perhaps not surprisingly then, theoretical efforts to comprehend firm competitiveness within regions are strong in their assertion that the key ingredients shaping firm competitiveness are predominantly endogenous to the region and reside within the environment within which businesses operate. Institutional theories focus on the role of various soft factors such as social capital or the norms and trust developed between businesses in a region. Supportive institutions are deemed to be particularly important and Amin and Thrift (1994), for example, have argued that 'institutional thickness' is the secret ingredient found in successful places and absent in others. By this is meant a complex of strong, viable and interacting institutions (both formal and informal) that share a sense of common enterprise. In a similar vein, cluster theories suggest that regions with strong localized clusters of specialized export-orientated businesses and inter-firm linkages will be more successful than others, since these will create the local external economies such as access to pools of specialized labour and knowledge that provide increasing returns to firms. Other theories, such as endogenous growth and neo-Schumpeterian approaches, focus much more specifically on innovation and entrepreneurship and their regional determinants as critical drivers for firm and regional competitive success. These theories argue that localized accumulations of skilled workers, entrepreneurship and innovation create powerful increasing returns for firms and create virtuous cycles of innovation, local spillovers and growth.

To this list may also be added a whole host of other qualitative or soft location factors which different theoretical contributions have variously suggested may create a favourable environment for the development of productive firms. These include the local entrepreneurial culture, forms of bank–industry relations, forms of labour–management relations, the quality of the local living or social environment, the cultural resources of a region, and the regional identity and international image (see Cambridge Econometrics et al., 2002 for a more detailed review). Since these soft location factors are also those which may help attract in new investment and labour, the blurring of boundaries between competitiveness and competition referred to above is plain to see.

This macroeconomic conceptualization thus asserts that the competitiveness of a region is deemed to reside both in the competitiveness of its constituent firms and their interactions, and in the wider social, economic, institutional and public attributes of the region itself. In short, 'the region' matters and fundamentally shapes the volume and rate at which human capital is employed as well as the ability to attract mobile capital and labour and thus, ultimately, the capacity to be efficient in high value-added growth sectors.

The coherence and utility of these theoretical contributions can, however, be questioned. They clearly do not constitute a single, comprehensive theory of regional competitiveness. Furthermore, there exists a distinct lack of clarity as to the precise significance of the different firm and region-centric factors deemed to influence it. Indeed, there has been a tendency to assert that the myriad of determinants of firm competitiveness uncovered by theoretically driven efforts to comprehend it are of equal importance in all spatial and locational contexts. Thus, Malecki (2002: p. 941) observes 'all of the issues that have risen to the top of the research agenda over the past 30 years are relevant – indeed, essential … Having only *some* of these conditions in good order is not enough'. Similarly, Deas and Giordano (2001) assert that the competitiveness literature has tended to proffer a generalized checklist of relevant determinants of firm competitiveness and that further empirical research is required to test the relative importance of factors deemed to generate competitive advantages for internationally successful firms.

Whilst region and localization economies are clearly important to the performance of firms, their precise role and significance remains somewhat vague. By way of illustration, in a paper asserting the role of the region ('territory') in shaping firm competitiveness, Camagni (2002) asserts that the companies and entrepreneurs that compete in international markets are 'to a large extent generated by the local context' (p. 2396). One might question what exactly constitutes 'a large extent' and whether this necessarily holds for all firms in all regions. Gardiner (2003: p. 20) argues for 'case studies of successful regions to assess the importance of more qualitative factors, e.g. regional governance, and the ability to transfer the factors driving success to those areas which are currently less competitive'. Others have noted that measuring the precise impact of institutional or collective factors such as untraded interdependencies and institutional thickness on regional innovation, productivity and competitive advantage has proved notoriously difficult empirically and results have been mixed (Cambridge Econometrics et al., 2002).

Indeed, there are various grounds for questioning both the universality and strength of the region–firm competitiveness nexus. In some instances the competitiveness of firms in the region may be altogether disconnected from the region. This is certainly the case with multinational enterprises (MNEs), the competitiveness of which may reflect conditions in the parent country as much as the host one. Even non-MNE firms find themselves more and more locked into, and therefore shaped by, the international networks in which they increasingly participate.

This is not to say that the region and localization economies are irrelevant to the performance of firms. Rather, the argument here is that their significance may perhaps have been overstated. In practice, the region's influence may vary, depending on the particular industrial structure and context, the balance of globally and locally oriented firms, and the degree to which the region constitutes an internally cohesive economic space with reasonably homogenous environmental assets. The critical point is that there has been limited research here and we actually know very little about the spatial scales over which key external or localization economies operate or whether they can be developed equally across all parts of a regional economic space (Kitson et al., 2004). What is also clear is that the theoretical contributions around this conceptualization of competitiveness focus overwhelmingly on business performance and say very little about employment and how this might shape overall performance.

Other authors dispute the notion that competitiveness can be directly equated to regional prosperity and instead argue that competitiveness is 'revealed' through the performance of the region's firms. Whilst this approach clearly differs from the macroeconomic conceptualization discussed above, it derives from the same output-based discourse of firm-based competition and instead sees regional competitiveness in microeconomic terms, as an aggregate of firm-based competitiveness or productivity.

According to Porter (1990; 1995), firm competitiveness is simply a proxy for productivity. Porter has argued that firms that are capable of producing more output with fewer units of input than their rivals generate a 'competitive advantage' (his preferred term) in the markets in which they compete, enabling them to grow and prosper accordingly. A firm's productivity, he argues, is dependent upon its 'entrepreneurialism'. This is defined as its capacity to innovate in the production process, to access new and distinctive markets in different and unconventional ways, and to produce new or redesigned goods and services with perceived customer benefit. Thus, according to

Porter, firm competitive advantage is not simply centred on a narrow efficiency-based conception of productivity, but also depends on the value of products and service produced, i.e. their uniqueness and quality.

Porter has extended and applied his model of the competitive advantage of firms to the competitive advantage of regions and nations and thus asserts that the productivity of a region (and indeed any territorial entity) is ultimately set by the productivity of its firms, whether they are domestic firms or subsidiaries of foreign companies (Porter, 1995; 1998). The essence of his argument is that 'comparative advantage is created and sustained through a highly localized process' (Porter, 1990: p. 19) and thus that firm productivity is critically shaped by the quality of the microeconomic business environment:

> More sophisticated company strategies require more highly skilled people, better information, improved infrastructure, better suppliers, more advanced research institutions, and stronger competitive pressures, among other things.
>
> (Porter, 2003: p. 25)

In his renowned diamond model, Porter identifies four sets of factors as critical elements of the microeconomic business environment, i.e. demand conditions; factor (input) conditions; the context for firm strategy and rivalry; and related and supporting industries. A region's relative competitiveness depends on the existence and degree of development of, and interaction between, these four key subsystems. In particular, functioning clusters of interrelated firms are critical. His argument is that strong business networks and regionally based relational assets provide a major benefit to firms' competitive performance by encouraging them to continually improve their operational effectiveness.

Porter has successfully branded, transformed and exported his diagnosis of regional competitive advantage to development agencies and governments all over the world, thus establishing his status as a key player in the global spread of the competitiveness discourse (see Chapter 2). However, his ideas have been criticized for lacking precision, determinacy and strong predictive ability (Grant, 1991). In particular, it is questionable whether this micro-level analysis of the competitive advantage of firms can be so easily applied as an explanation of the macro-level performance and development of regions. Porter simply presumes some 'invisible hand' whereby the pursuit of competitive advantage by firms translates into increasing productivity and prosperity for the regional economy as a whole, leaving unanswered many questions as to how this happens.

Others have sought to flesh out the links between the productivity or efficiency of a region's firms and its overall prosperity. Camagni (2002) asserts that regions compete in terms of absolute advantage or efficiency, not comparative advantage. A region may be thought of as having absolute competitive advantages when it possesses superior technological, social, infrastructural or institutional assets that are external to, but which benefit individual firms and give them higher productivity. Similarly, and in an apparent change in his thinking, Krugman (2003) has argued that a region that is more efficient (productive) will be able to attract (and retain) labour and capital from other regions. Furthermore, he suggests that these factor inflows will subsequently reinforce the region's absolute productivity advantage even further in a virtuous circle.

However, the conceptualization of regional productivity remains somewhat vague. Labour productivity (output per unit of labour) coupled with the employment rate represent what might be termed measures of 'revealed competitiveness', in that they are both central components of a region's economic performance and its prosperity, but of themselves say little about the underlying attributes and complex processes on which they depend. Indeed, there is considerable confusion as to what actually causes some regions to have higher productivity than others. Different theoretical approaches place emphasis on different factors, with neoclassical growth theory highlighting the role of different factor endowments, and especially differences in labour/capital ratios and technology, endogenous growth models focusing on differences in the knowledge base and proportion of the workforce in knowledge-producing industries, and new economic geography approaches focusing on the increasing returns that give local firms higher productivity (Gardiner et al., 2004).

The evolutionary perspective appears to offer some promise in respect of its ability to explain why some regions may develop firms with a competitive advantage (Boschma, 2004). The evolutionary approach emphasizes dynamic competitive advantage and highlights the ability of a regional economy to adapt to changing market conditions and the emergence of new technologies and competitors. It asserts that a region's competitive advantage is a product of its historical, path-dependent development and its capacity to create new development trajectories. Each region is thus acknowledged to be unique and to be capable of influencing its competitive environment through its institutional forms, their ability to adapt and learn and thus provide a positive stimulus for regional growth. This perspective is still, however, in its infancy and requires further empirical testing and possible refinement.

The policy discourse

This discussion illustrates that regional competitiveness lacks a clear, unequivocal and agreed meaning within the academic literature and there is no clear theoretical or conceptual framework for understanding regional competitive performance. It is perhaps not surprising therefore that the policy discourse around it is equally, or not more, muddled and confused.

In general, policy makers have tended to favour the macroeconomic definition of competitiveness which directly equates it with regional prosperity. Thus, for example, the UK government states that regional competitiveness 'describes the ability of regions to generate high income and employment levels whilst remaining exposed to domestic and international competition'

(DTI, 2003: p. 3), a definition also utilized by the European Commission (CEC, 1999a: p. 75). Policy makers also appear strongly persuaded of the importance of factors endogenous to the region's microeconomic business environment in shaping firm competitiveness. The European Commission (CEC, 2004) asserts that developing regional competitiveness depends on 'encouraging the development of knowledge-based economic activities and innovation' (p. 3), and observes that 'there is a growing consensus about the importance for regional competitiveness of good governance – in the sense of efficient institutions, productive relationships between the various actors involved in the development process and positive attitudes towards business and enterprise' (p. xiii). Similarly, in 2001 the UK government announced that it was introducing a new approach to regional policy which was intended to strengthen the capability of regions to build on their own competitive advantages by 'boosting regional capacity for innovation, enterprise and skills development' (DTI, 2001: p. 4). Furthermore, all existing attempts to define and measure regional competitiveness internationally incorporate indicators relating to the quality of the business environment, business density and clustering and/or knowledge intensity and innovation, albeit in different combinations. The nature of these indicators and their inevitable use in regional benchmarking also help to perpetuate the notion that regions are engaged in fierce and direct competition with one another over these key resources and in respect of outcomes (see Chapter 4).

To further confuse matters, regional productivity has also begun to assert itself in the policy lexicon. In the UK, the Treasury has identified productivity as the key to achieving higher living standards, and has asserted that differences in regional income levels (as measured by GDP per capita) are a function of variations in regional productivity and the employment rate (HM Treasury, 2001). Five 'drivers' of productivity are identified: skills, investment, enterprise, innovation and competition. These drivers clearly do not correspond directly to any of the theoretical contributions discussed above and appear to have emerged instead from more practical models of national economic growth. As Fothergill (2005: p. 661) observes, 'whether a framework developed to understand national differences in growth translates well to the regional scale is of course something that needs to be questioned. Whether the five drivers of productivity provide the basis for an effective practice strategy, as opposed to just a handy framework for thinking about issues, is also unclear.' Not least amongst Fothergill's concerns is the failure of this policy statement to acknowledge that differences in regional productivity rates will, in part at least, reflect differences in industrial structures.

The UK DTI, which regularly publishes regional competitiveness indicators, is more equivocal in its approach and shies away from identifying explicit causal relationships between productivity and living standards. Instead, it publishes a series of different indicators of, *inter alia*, regional prosperity, productivity, infrastructure and labour market performance which are deemed to represent 'a balanced picture of all the statistical information relevant to regional competitiveness' (DTI, 2003: p. 3). No explanation is provided as to how, or to what extent, the various indicators relate to the overall definition of regional competitiveness (see also Chapter 4).

This reveals the existence of considerable confusion within policy circles as to whether a region's competitiveness is simply reflected in its overall levels of prosperity, or whether the conditions shaping the region's ability to sustain its macroeconomic performance are more or less important considerations. The outcome is a conception of regional competitiveness that is a chaotic mix of productivity, prosperity and the development of numerous different attributes of the business environment. The policy discourse around competitiveness clearly tends to conflate the strategic pursuit of 'competitiveness' as meaning the search for improved economic performance, with engagement in competition for resources and the development of boosterist strategies aimed at attracting high-quality, innovative, 'knowledge-based' firms and skilled labour. Competitiveness is a catch-all for the pursuit of business-led growth and entrepreneurial place selling.

Conclusions

This chapter has examined the existing literature on the concept of regional competitiveness and has sought to critically unpack its meaning in academic and policy debates as well as its roots in theory. This has demonstrated that whilst the notion of regional competitiveness is firmly ensconced in, and clearly shaping the broad direction of, regional economic development policy, it is a rather chaotic discourse. For some, regional competitiveness is used to describe the competitive behaviour of regions as they seek to attract the resources and inputs critical to securing wider economic success. Yet this assumes regions are actors with discrete control over their jurisdictions and that they are actively in competition with other, similarly constituted regions in clearly defined markets, something which remains subject to debate. Regional rivalry clearly does exist, but this does not necessarily lead to clearly defined performance outcomes as it does for firms, and thus it becomes important to distinguish between competition and competitiveness. Regional competitiveness is more clearly intended to refer to the relative economic prosperity of a region, although again there is considerable confusion as to what this actually means and when a situation of 'competitiveness' has indeed been achieved.

There are fundamental problems in equating competitiveness directly with macroeconomic performance or living standards in a region, not least because of the vagueness which surrounds the conditions that are required to ensure regions with productive firms can generate the high profits and wages critical to enabling them to sustain prosperous economies. Whilst there may be more justification for defining competitiveness in terms of productivity, and especially a region's comparative advantage, productivity is a complex, 'revealed' measure which conceals more than it reveals about the determinants and nature of underlying regional economic development processes.

One of the main problems is, as Kitson et al. (2004: p. 997) observe, that 'we are far from any agreed framework for defining, theorizing and empirically analysing regional competitive advantage', although they themselves, along with other authors such as Budd and Hirmis (2004), make some progress towards synthesizing from available literatures the different factors that need to be included. Instead, there is a range of disparate bodies of work, each of which tends to adopt a different perspective and to emphasize a different set of key variables. However, it is apparent that there clearly are some dominant axioms which collectively define the discourse, notably that regional competitiveness is a firm-based, output-related conception strongly shaped by the microeconomic business environment. Indeed, there is a clear view that the competitiveness of a region resides in the competitiveness of its firms and so-called 'regional externalities' or the resources that reside outside individual firms but are drawn on by those firms and which lie within the wider social, institutional and economic assets of the region itself (Kitson et al., 2004).

Given this context, it is perhaps not surprising, then, that the policy discourse on competitiveness tends to conflate the strategic pursuit of competitiveness with engagement in competition for resources, and tends to identify a hotch-potch of various different determinants of regional competitive advantage as being critical to enhanced performance. The discourse of regional competitiveness perhaps not surprisingly appears to exemplify a wider tendency towards 'theory led by policy' (Lovering, 1999), whereby commentators have found themselves engaged in *ex post* rationalization of a term that has already become embedded in common policy parlance.

This begs numerous questions about the political economy of the discourse – questions which, to date, have been given much less attention in the relevant literatures. These include how and why a discourse with ostensibly thin and ill-defined content has assumed such significance in regional policy circles; who decides which particular definition and meaning of regional competitiveness is adopted in different contexts and why; what determinants and strategies are deemed to be critical to success; what role regions themselves play in these processes relative to the roles of national states and supranational institutions; and what difference competitiveness strategies or policy interventions make, if any, to regional economic outcomes. These form key questions to be explored in the following chapters.

2 The political economy of regional competitiveness

Introduction: competitiveness and political economy

The concept of competitiveness is demonstrably a chaotic one, yet undeniably pervasive, and has become particularly prominent for regions. Indeed, according to Lagendijk (2007: p. 1201), competitiveness is one of a number of key ideas and issues that have become 'sutured into the canvass [*sic*] of the region'. Competitiveness is an exemplary neoliberal discourse which has acquired 'hegemonic' status in that it is a central system of practice and value 'which we can properly call dominant and effective' (Williams, 2005: p. 38). This begs the question as to how and why a concept with ostensibly confused, narrow and ill-defined content has become so salient in regional economic development policy and practice as to constitute 'the only valid currency of argument' (Schoenberger, 1998: p. 12).

This chapter will argue that critical to understanding the competitiveness hegemony is developing an understanding of the policy process, which refers to all aspects involved in the provision of policy direction for the work of the public sector. This therefore includes 'the ideas which inform policy conception, the talk and work which goes into providing the formulation of policy directions, and all the talk, work and collaboration which goes into translating these into practice' (Yeatman, 1998: p. 9). A major debate exists in the policy studies literature about the scope and limitations of reason, analysis and intelligence in policy making – a debate which has been reignited with the recent emphasis upon evidence-based policy making (see Davies et al., 2000). Keynes is often cited as the main proponent of the importance of ideas in policy making, since he argued that policy making should be informed by knowledge, truth, reason and facts (Keynes, 1971, vol. xxi, 289). However, Majone (1989) has significantly challenged the assumption that policy makers engage in a purely objective, rational and technical assessment of policy alternatives. He has argued that, in practice, policy makers use theory, knowledge and evidence selectively to justify policy choices which are heavily based on value judgements. It is thus persuasion (through rhetoric, argument, advocacy and their institutionalization) that is the key to the policy process, not the logical correctness or accuracy of theory or data. In other words, it is interests as much as ideas that shape policy making in practice.

This connects to much longer and wider intellectual debates in the history of economic thought about the role of interest in motivating individual or group behaviour, with Hirschman (1977) playing a prominent role in asserting that the ideological foundations of capitalism reside in the general phenomenon of violent passion being subdued by innocuous interest in acquiring wealth. Indeed, Hirschman (1986) observes that 'interest' or 'interests' is one of the most central and controversial concepts in economics and more generally in social science. Since coming into widespread use since around the latter part of the sixteenth century, the concept has stood for 'the fundamental forces, based on the driver for self-preservation and self-aggrandizement, that motivate or should motivate the actions of the prince or the state, of the individual and, later, of groups of people occupying a similar social or economic position (classes, interest groups)' (Hirschman, 1986; p. 35).

Recent developments in political economy and, in particular, the emergence of discursive and interpretative approaches to understanding policy, provide a potential framework for understanding why and how particular ideas, such as competitiveness, arise, spread and become hegemonic. The value of such approaches lies ostensibly in their ability to help avoid naive acceptance of such ideas as economic phenomena, and instead to provoke interrogation of the political dynamics of their evolution and their utility for advancing particular policy goals and discouraging others.

The purpose of this chapter is thus to examine the international origins and trajectory of the regional competitiveness discourse. This focuses primarily on the rhetorical power and usefulness of the discourse for particular groups with particular interests, but also highlights the complex coming together of a range of different economic, institutional, pragmatic and political variables which collectively reinforce the hegemony of the discourse.

The chapter begins by exploring the nascent Cultural Policy Economy approach to analysing the evolution of key discourses such as competitiveness. This is identified as a potentially useful framework for examining a number of critical, unanswered questions around the process of rescaling competitiveness to the region. It then proceeds to utilize this approach to trace the origins and trajectory of the regional competitiveness discourse. This explores how and why an ostensibly national policy discourse was rescaled to the regional level, and considers the implications of this both for the competitiveness discourse and for regions themselves.

Cultural political economy

A number of studies have applied political economy approaches to enhance our understanding of the status and role of global competitiveness. Rosamund (2002), for example, demonstrates that in the context of Europe, competitiveness is a powerful discursive construction that has become 'banal' inasmuch as it is accepted as commonsensical and thus not worthy of critical discussion. This, he argues, serves as the basis for asserting the sense of a

European economic space and thus for enhancing the legitimacy of both the 'Europeanization' of governance capacity and the deepening of integration processes. Cammack (2006) focuses on the spread of the global competitiveness discourse and argues that this is a deliberate tactic on the part of developed countries and bodies such as the OECD (which he labels the 'convergence club') to maintain and reproduce the capitalist hegemony within advanced capitalist countries themselves. Similarly, Fougner (2006) asserts that the notion of international competitiveness is deliberately constituted and follows directly from contemporary efforts to govern the world economy in accordance with a neoliberal rationality of governance. This in turn privileges a particular 'attractiveness' oriented conception of competitiveness (see Chapter 1) which constitutes states and their populations as 'competitive and entrepreneurial "place-sellers" in a global market place for investment' (p. 181).

These studies may, however, be criticized for generating a notion that competitiveness is ostensibly a global construct transferred in a rather linear and unproblematic fashion to other sites and scales (including regions) by powerful global actors utilizing specific techniques and networks. The emerging body of work becoming known as Cultural Political Economy (CPE) suggests that a more complex transmission mechanism may be at work.

CPE has been developed by Jessop (2005) and Sum and Jessop (2001) in an effort to understand the contribution of discourse and discursive practices to the forming of subjects, and how they come to be naturalized and materially implicated in everyday life, for example through policy choices. This builds on the strategic-relational approach (SRA) initially developed by Jessop (2001), which has followed an evolutionary political economy approach to exploring patterns of accumulation, regulation and governance and thus the role and primacy of state strategy and political projects. CPE holds that technical and economic objects are always socially constructed, historically specific and in need of continuing social 'repair' work for their reproduction. In this way, CPE emphasizes the complex relations between meanings and practices and seeks to trace the co-evolution of intersubjective meanings (semiosis) and extra-semiotic processes in securing institutional stability, incorporating agency in socio-economic management, and in selecting and retaining particular discourses for state intervention.

More specifically, CPE emerged from research into the 'entrepreneurial city' which was instrumental in the development of two key notions. The first of these was the assertion that cities or regions are actors or 'spaces for themselves' endowed with 'capacities to realize particular discursive-material accumulation strategies and hegemonic projects' (Jessop and Sum, 2000: p. 2310). This was significant in its direct implication that cities and regions are capable of pursuing innovative strategies to maintain or enhance their economic competitiveness vis-à-vis other economic spaces. As such, they are increasingly like nations with the potential to act as 'competition states' (as Cerny, 2007). The second notion was that of 'economic imaginaries'. These are subsets of economic activity (such as competitiveness, innovation and workfare) which are discursive constructs transformed into sites and objects of observation, calculation and governance and thus used by the state to secure hegemony (Sum and Jessop, 2001).

A number of studies have applied the CPE framework and drawn attention to its utility. For example, Burfitt et al. (2006) and Jessop (2005; 2008) have applied CPE to the knowledge-based economy which has emerged as a key narrative of state policy in western democracies since the mid 1990s. These analyses have served to highlight how the discourse of the knowledge-based economy is narrated and disseminated internationally as an opportunity, a threat and a challenge, all of which require and legitimate policy responses. Similarly, Mulderrig (2008), in an exploration of changes in educational discourse in the UK, has demonstrated how national competitiveness is used as a catch-all and cure-all concept to justify the imperative of economic growth. This illustrates that such discourses are constructed to sustain and support the growing neoliberal hegemony and to legitimize the extension of certain strategic and spatial policy activities.

The CPE approach posits that the construction of an economic imaginary involves a number of distinct evolutionary stages. First, there is a process of 'selection' in which particular discourses are prioritized in terms of their ability to interpret and explain particular events. Second, is the process of 'retention' in which these discourses are repeatedly incorporated into as many institutional sites, roles, routines and strategies as is possible. Third, is the process of 'reinforcement' whereby these discourses are restated and embedded in procedural mechanisms, governance structures, rules and regulations such that they become a naturalized discourse (Jessop, 2005). According to Jessop (2005) there are a range of institutions and actors such as government departments, business organizations, the OECD, the EU and the World Bank that are likely to seek to establish and institutionalize such imaginaries. They use a range of discursive vehicles or 'genre chains' (Sum, 2004) such as policy documents and strategy reports to reinforce these imaginaries to suit their particular purposes.

However, whilst economic imaginaries can be discursively constructed and materially reproduced at different sites, on different scales and with different spatial and temporal horizons, they are only ever likely to be partially constituted and will remain contingent. As Jessop (2005: p. 146) observes, the process of material reproduction 'always occurs in and through struggles conducted by specific agents, typically involves the asymmetrical manipulation of power and knowledge, and is liable to contestation and resistance'. This means that there will always remain space for competing imaginaries to challenge the dominant ones.

Indeed, by making reference to the contributions of various other authors, it is possible to suggest that a fourth evolutionary phase in the development of economic imaginaries (particularly competitiveness) may be conceived – a process which may be labelled 'recontextualization'. Milliken (1999) asserts that actors are constantly involved in a struggle over competing discourses (or narratives) that threaten the hegemony of existing logics and rationales for action. This struggle, she argues, can ultimately change narratives, although she does not elaborate precisely how or why this occurs. Prytherch (2006: p. 204) is more specific and, using a case study of the emerging development narrative in Valencia in Spain, argues that regions and their governing institutions negotiate and articulate the forces of global change more locally to make competitiveness 'particularly regional' and attuned to their specific agendas (a similar argument is posited by Rodaki in relation to Rome; Rodaki, 2008).

Similarly, in a study of urban policy in the northern Spanish city of Bilbao, Gonzales (2006) examines how local policy actors actively transform and redefine dominant discourses about the contemporary rescaling of the capitalist economy in order to suit local conditions and justify an entrepreneurial policy approach. Gonzales (2006: p. 839) argues that scale itself is discursively constructed and develops the concept of 'scalar narratives' which she defines as 'the stories that actors tell about the changes in the scalar localization of socio-political processes'. These discursive strategies are employed to give coherence to their scalar political practices and to legitimate and justify a particular approach to urban policy that focuses on economic competitiveness. This raises interesting questions as to whether the discursive constructions of scale and of competitiveness are in fact distinct, or whether and to what extent their evolutionary processes are inherently interrelated and complementary.

Lagendijk (2007) provides some important insights here. In a paper examining the emergence of the region as an 'omnipresent' imaginary, Lagendijk highlights the complex state and strategic selectivities that led the region to become an 'available window to experiment with new regulatory forms and "fixes"' (p. 1204). Whilst the discourse of the region has been strongly orientated towards economic aspects of regional development, Lagendijk observes that 'this does not mean that economic development, notably its neoliberal connotation of "competitiveness" is always framed as the primary goal or condition' (p. 1204). The privileging of the region instead reflects both economic and non-economic narratives such that regions are continuously subjected to multiple spatial and scalar selectivities of the state. Indeed, Lagendijk contends that securing the coherence of these often conflicting narratives within the constraints afforded by the particular exigencies of regional policy space presents a major and enduring challenge for effective strategic policy action by regional actors (see also Chapter 5).

This, in turn, points to the need to develop and extend the CPE approach to allow a better understanding of the contemporary political economy of the state. This is one of the central arguments developed by Jones (2008: p. 386), who, in a very thorough exposition of CPE's strengths and weaknesses, articulates the need for 'a more systematic and sophisticated understanding of the reshaping of state spatiality (or state/space) *in tandem with* the semiotic turn implied by CPE', thus implying that processes of state rescaling and the spread of key economic imaginaries are related to, and help to explain, each other.

Jones goes on to propose two key ways in which CPE can be modified and thus extended to develop its geographical dimension and sited complexities and contradictions. The first is to explore the geography of discourse or how ideas, concepts and theories are produced, disseminated and consumed. Key to this geography, he argues, is the construction of 'discursive sites' such as research centres and policy fora, out of which new institutional forms and cadres of experts emerge. The second is to focus on the play of binaries and to develop an understanding of how key discursive constructs are always contextualized in relation to 'unstated' concepts and ideas. This, he asserts, will help to elucidate the relative importance of material and non-material (semiotic) factors in shaping how successfully the imaginary will be retained and reinforced. He concludes by reflecting on the value of a modified CPE approach, arguing that 'it is not the components of cultural political economy that should make this perspective distinctive, *but the way the components made visible by CPE interact and relate to each other*' (p. 393). Whilst acknowledging that the modified CPE approach requires more detailed empirical application and testing, Jones illustrates its potential through a study of the skills society in the UK which highlights, for example, how key policy statements on the knowledge-based economy and welfare-to-work agendas work in tandem to advocate a work-first approach to tackling unemployment.

Hudson (2008) has also offered some interesting critical reflections on the CPE approach. In a paper exploring the value of the CPE approach to the analysis of global production networks, Hudson states that as yet there has been little serious engagement with the materiality of the economy and so with the relations between the material, semiotic and political economic. He goes on to argue that emphasizing the materiality of the economy highlights the value lost to capital through waste and thus 'provides a perspective from which to review the ecological implications of the repertoire of social choices about how and what to produce, exchange and consume and where to do so' (Hudson, 2008: p. 437). Harding (2007) makes a similar plea for reinserting the 'economic' into 'political economy'.

The discourse of regional competitiveness presents a very rich and fertile context for applying and possibly developing the CPE framework, since there are numerous unanswered questions surrounding its dominant position. Lagendijk has articulated some general questions about the spread of key discursive constructs to regions, which are certainly pertinent to the issue of regional competitiveness (p. 1201): 'Who and what mediated the selection? What kind of metaphors, knowledges and technologies become dominant? What presentations of "problems" and "solutions" are adopted in policy processes? How are policy outcomes and associated territorial developments monitored and relayed? And to what extent are these aspects differentiated in space and time?'

But other questions also emerge from the discussion in the previous chapters of this book and in the above review of CPE. These include: How does the evolution of the scalar narrative of the region, and 'New Regionalism' in particular, relate to and complement the evolution of the competitiveness discourse? Which definition of regional competitiveness is used to shape strategies? How is this decided and by whom? Are regions putative agents (subjects) of competitiveness able to shape, mould and possibly resist its imperative, or simply objects of competitiveness policy? What role, if any, is played by counter-hegemonic forces in generating agonistic debate and struggle between actors within regions and thus challenging competitiveness projects and strategies? How are these different forces continually balanced and counterbalanced in 'an unstable equilibrium of compromise' (Jessop, 2005: p. 161) within regions? What role do binaries play in how these discursive contests are played out? What role does the competitiveness discourse play in shaping policy in regions? Is it a banal, taken-for-granted proposition that provides a weak backdrop to policy, or a rather more powerful concept utilized more knowingly as a strategic device? What role, if any, do material considerations provide in influencing the nature of competitiveness strategies, their scope, effectiveness and limitations?

Answering these questions (in this and subsequent chapters) will help to create an understanding of how and why a particular conception of 'regional competitiveness' has developed and taken hold, as well as provide the foundation for understanding the scope for its recontextualization and possible contestation in different regional contexts. This will not only allow for further reflection on the utility of the CPE approach, but also provide a basis for thinking more critically in later chapters about the need and potential for more diverse, context-specific and transformatory policy agendas in regions.

The discourse of national competitiveness: origins and selection

Whilst it is difficult to date its origins precisely, the discourse of national competitiveness is historically and to a large extent also geographically specific, having become prominent in the United States in the mid to late 1980s as a direct antecedent of the globalization discourse. The idea of competitiveness proved increasingly commonplace in political discussions regarding the dilemmas facing the American economy in the context of the breakdown of the post-war settlement and the rise of the neoliberal economy (see Krugman, 1997b).

The notion of national competitiveness quickly became prominent. Led by specific business sectors and companies concerned with the rising economic power of nations such as Germany and Japan, national competitiveness became constituted as a governmental problem and conceived in stark competition-based and world market terms. Thus, in a classic exposition of the emerging thinking, Thurow (1992) asserted that the structural properties of the global economy had changed, creating an intense world of direct 'head-to-head' competition between nations for mobile capital and investment, and the seemingly inevitable consequence that whilst some would win, others would lose.

Furthermore, a particular conception of national competitiveness quickly became prevalent. This went beyond conceiving it as the mere sum of the competitiveness of national firms, to include a concern with living standards (Fougner, 2006). Indeed, in what became the standard definition of competitiveness at the time, Scott and Lodge defined national competitiveness in 1985 as 'a nation's ability to produce, distribute and service goods in the international economy in competition with goods and services produced in other countries, and to do so in a way that earns a rising standard of living. The ultimate measure of success is not a "favourable" balance of trade, a positive current account, or an increase in foreign exchange revenues. It is an increase in the standard of living' (Scott and Lodge, 1985: pp. 14–15). Competitiveness was thus more than just an end in itself – it was construed as a powerful *means* to achieve the ultimate end of rising prosperity. Grow and attract the right sort of investment and the most globally competitive firms and, according to the emerging discourse, a nation would enjoy greater prosperity. In the early 1990s this definition of national competitiveness was subsequently adopted by the OECD as a key policy goal, and a suite of 'competitiveness policies' focused on free trade, deregulated labour markets and investment in human capital established as priorities for its member nations (OECD, 1992). Similarly, in 1995 UNCTAD (the United Nations Conference on Trade and Development) asserted the importance of national competitiveness and its conception as something akin to national welfare and certainly beyond a simple linear aggregation of the competitiveness of individual firms (UNCTAD, 1995). The International Institute for Management Development (IMD) adopted an even broader, more complex conceptualization defining competitiveness as 'the ability of a country to create added value and thus increase national wealth by managing assets and processes, attractiveness and aggressiveness, globality and proximity, and by integrating these relationships into an economic and social model' (IMD, 1996: p. 42).

'Competitiveness-speak' spread quickly across Western Europe and, from the mid 1980s, became central to the discussions around the integration of Europe's national economies. The notion of the 'competitive threat' posed by the economies of the US and East Asia provided a key rationale for market liberalization across the European Community, and ultimately helped to propel the Single Market project (Rosamund, 2002). The year 1993 saw the publication of the European Union's pivotal White Paper on Growth, Competitiveness and Employment, which drew on the Porterian notion of competitive advantage and established that the principal policy responsibility for the EU and its national authorities was to 'provide industry with a favourable environment, to open up clear and reliable prospects for it and to promote its international competitiveness' (CEC, 1993: p. 5).

Competitiveness became progressively more embedded in the EU's constitution, over time forming an important backdrop to the Maastricht (1992) and Amsterdam (1999) treaties which focused on deepening the integration process. By the year 2000 and the agreement of the Lisbon agenda, the pursuit of competitiveness was firmly established as the EU's principal strategic objective with the 10-year development plan aimed at making the EU 'the most dynamic and competitive knowledge-based economy in the world capable of sustainable economic growth with more and better jobs and greater social cohesion' (Lisbon European Council, 2000). Significantly, the meaning of competitiveness was not precisely defined. Nevertheless, the assumptions underlying its use were clearly revealed in the surrounding policy discourse: namely that the global economy is increasingly a competitive place; that the world economy is increasingly interconnected such that what happens in one economy invariably and inevitably directly affects others; that nations as well as firms compete and so it is necessary for countries to improve their competitiveness; and that the European economy suffers from a 'failure of competitiveness' characterized by overly high labour costs in manufacturing and a lack of entrepreneurship (Turner, 2001).

Within the EU, the pursuit of national competitiveness assumed particular prominence within the UK. The Conservative government created a Competitiveness Unit in 1993 and in successive White Papers subsequently adopted and prioritized the prevailing macroeconomic notion of national competitiveness. However, as Mulderrig (2008) demonstrates, the notion of competitiveness was not simply translated from the EU to the UK. Rather, it was recontextualized and moulded to fit the particular socio-historical and political context of the UK. In particular, competitiveness was used in vague, polyvalent terms as a panacea, and of widespread necessity, benefit and relevance to the ultimate project of economic growth. Furthermore, a notable feature of the 'project' of *competitiveness* appears to be the strategic use of managerial instruments in an attempt to define its determining factors (Mulderrig, 2008). The competitiveness theme continued under the New Labour government elected in 1997, with the establishment of a Competitiveness Council and the publication of further White Papers in which it was a dominant theme. Under Blair, however, competitiveness was articulated around skills, with active labour market polities deemed to be key to fulfilling competitiveness strategies.

The concern with national competitiveness has more recently spread beyond the US and Western Europe. It has now reached the shores of Japan, China, South America and indeed much of the rest of the world, including Central and Eastern European countries and developing countries (see, for example, UNCTAD, 2004; Drahokoupil, 2008), such that there now exists an international policy consensus on the importance of competitiveness and the policy approaches deemed to enhance it.

Retention and reinforcement

The universal acceptance of the discourse of national competitiveness reflects the power of the key interests, institutions and networks that have promoted and disseminated its logic. Ultimately, the language of competitiveness is the language of the business community. Thus, critical to understanding the power of the discourse is, firstly, understanding the appeal and significance of the discourse to business interests and, secondly, exploring their role in influencing the ideas of policy elites who then reinforce and spread the competitiveness message.

Part of the allure of the discourse of competitiveness for the business community is its seeming comprehensibility. Business leaders feel that they already understand the basics of what competitiveness means and thus it offers them the gain of apparent sophistication without the pain of grasping something complex and new. Furthermore, competitive images are exciting and their accoutrements of 'battles', 'wars' and 'races' have an intuitive appeal to businesses familiar with the cycle of growth, survival and sometimes collapse (Krugman, 1997b). The climate of globalization and the turn towards neoliberal, capitalist forms of regulation have empowered business interests and created a demand for new concepts and models of development offering guidance on how economies can innovate and prosper in the face of increasing competition for investment and resources. Global policy elites of governmental and corporate institutions, who share the same neoliberal consensus, have played a critical role in promoting both the discourse of national and regional competitiveness, and of competitiveness policies which they think are good for them (such as supportive institutions and funding for research and development agendas).

Thus in the US in the 1980s the concerns of businesses and industry representatives prompted a growing number of business analysts and experts to develop corporate 'how-to-compete' models.[1] Most notable amongst these was Michael Porter, who developed a number of diagnostic tools for analysing the competitiveness of influential US firms. The interest in competitiveness tools and strategies quickly spread to nations and Porter himself adapted and extended his analysis to the sources of competitiveness at the national and regional level (Porter, 1990; see also Chapter 1). Porter's models were quickly picked up by influential business strategy consultants such as McKinsey and international organizations such as the World Economic Forum, and spread to businesses and governments worldwide. Porter himself played a powerful role in disseminating his ideas and applied the diamond model to countries as diverse as the USA, Portugal and New Zealand, whilst some of his colleagues further promoted the application of competitiveness analyses to developing countries in Latin America (Fairbanks and Lindsay, 1997).

Business interests also played a powerful role in propelling the OECD's adoption of the competitiveness imperative. Business demands for international benchmarking of the parameters of competitiveness (see Gassmann, 1994) played a powerful role in promoting the OECD to assert the importance of national competitiveness (OECD, 1992). In so doing, the OECD has played a key role in spreading the discourse through global policy networks and pushing a competitiveness-orientated policy agenda on to national governments in the industrialized world (Peet, 2003; Cammack, 2006). Indeed, the OECD has actively constituted itself as a 'convergence club' with an explicit focus on disseminating the imperative of global competitiveness to as many developed and developing countries as possible 'because it is the key to the maintenance and reproduction of capitalist hegemony within the advanced capitalist countries themselves' (Cammack, 2006: p. 13). This 'politics of convergence' has been developed through a range of mechanisms including the development of an internationally accepted set of appropriate policy choices seen as generative of capitalist competitiveness; the use of systematic surveillance, benchmarking and peer review to promote these policies to national governments; the identification of national governments as key agents of change; and the insistence that national governments shape public opinion to the logic of national competitiveness.

This activity has been subsequently supported and reinforced by a range of other international governmental organizations such as the World Bank 'whose policy guidelines are often framed in terms of the need of states to compete for footloose capital' (Fougner, 2006: p. 181), and a range of institutions such as the World Economic Forum and International Institute for Management Development, whose competitiveness indices serve as benchmarks of the relative ability of nations to attract footloose capital (see Chapter 4). Furthermore, numerous countries, such as the USA, Ireland and Singapore, have witnessed the formation of government-supported competitiveness councils that are engaged in understanding and promoting themselves as competitive places in the global market for investment.

Business interests also played a powerful role in asserting the importance of competitiveness and its pursuit through policy in the EU. The business lobby quickly became a very organized and effective grouping in Europe. The European Roundtable of Industrialists (ERT) – a forum of some 45 captains of industry from European multinational corporations – was established in 1983. It is essentially 'a club of individuals' (Balanya et al., 2000: p. 26), all of whom are highly influential industrialists with established access to both national and European decision makers.

Strengthening European competitiveness within the global economy has always been the ERT's main objective and it has consistently lobbied for the promotion of Europe-wide competition and competitiveness (see ERT, 1994; also Mazey and Richardson, 1997). The power of this lobby gained added momentum in 1993 with the publication of a report by the Union of Industrial and Employers' Confederations of Europe (UNICE) explaining the extent of the decline in the competitiveness of Europe (UNICE, 1993). This was followed up with further reports in 1994 and 1997 which contained a range of proposed policy options including more flexible labour markets, reduced minimum wage levels, a smaller and more efficient public sector and an improved climate for entrepreneurs (UNICE, 1994; 1997). Business interests in the UK soon followed suit and by 1996 the Confederation of British Industry (CBI) had refocused its policy programme to a concern with competing in the global economy.

By 1995 the business lobby in Europe had succeeded in ensuring that pursuit of competitiveness had become institutionalized in EU decision making with the establishment of the Competitiveness Advisory Group (CAG). CAG was established by the European Commission President in February 1995 in pursuance of a recommendation of the Essen European Council (December 1994). CAG consists of six business representatives, three trade union representatives, two academics and various others. Its mandate is to produce a six-monthly report to the Commission on the state of the EU's competitiveness and to advise on economic policy priorities for stimulating competitiveness, growth and employment. Following the completion of the two-year term of the Group and in light of its avowed success, President Santer decided (in June 1997) to appoint a second Group (CAG II) with the same mandate.

Whilst CAG presents itself as an independent think-tank, in practice it has very close links in terms of personnel with the ERT which has, in effect, allowed chief executives from the ERT to present their recommendations through a formal body with official EU status. As well as pressing home the overwhelming importance of the competitiveness message, CAG has also played a powerful role in shaping the competitiveness discourse in Europe in two other key respects. Firstly, it has placed strong emphasis on the competitiveness of places, asserting that 'it is competitiveness at the level of the economy – region, country or supranational entity – which most arouses the interest of the public' (Jacquemin and Pench, 1997: p. 5).[2] Secondly, it sees its mission as promoting the general importance of competitiveness and policy approaches towards its achievement rather than entering more specific debates on the measurement of competitiveness and the ranking of competitiveness performance.[3] In this regard, it has helped to turn competitiveness into a rather vague and malleable policy 'garbage-can' (Mazey and Richardson, 1997) into which various problems and solutions can be pitched without having necessarily clear or predictable outcomes. Consequently, and as indicated earlier, competitiveness has become de facto the main goal of EU policy making, overriding all other concerns (Balanya et al., 2000). As such, the EU has played an important role in asserting itself as a space of competitiveness, and also as a key actor in disseminating the competitiveness discourse both outside its borders (for example, through efforts to secure the integration of the former socialist countries of Central and Eastern Europe into its Single Market project) and within its borders and, in particular, to its regions.

Reinforcement, rescaling and the beginnings of recontextualization

The notion that competitiveness was a concept that could also be applied to regions appears to have emerged first in the context of Europe in the late 1980s, but became a widely accepted notion over the course of the 1990s. By the late 1980s, the European Commission had established a research interest in the factors shaping the competitiveness of the regional business environment from the perspective of mobile or potentially mobile business interests. Thus research conducted for the European Commission in 1990 surveyed 9,000 business enterprises and identified 37 factors shaping regional competitiveness (IFO, 1990). Similarly, a study for the Commission in 1992 focused on mobile investors in manufacturing and identified 22 location factors which shaped the competitiveness of regions (NEI, 1992). At the same time, regional development programmes in the EU developed a sharp focus on business competitiveness within regions. This included programmes such as the European Strategic Programme for Research and Development on Information Technologies (ESPRIT), whose main goals were to promote intra-European industrial cooperation through research and development and to thereby furnish European industry with the basic technologies that it needed to bolster its competitiveness through the 1990s (Mytelka and Smith, 2002). By 1997, Eurostat had identified a range of over 50 indicators that could be used to measure the competitive performance of regions, with indicators covering the business environment as well as economic output and performance indicators, whilst the 2003 Competitiveness report featured a review of the meaning of regional competitiveness and the state of knowledge on key factors determining it (CEC, 2003). By 2004, regional competitiveness had been firmly adopted as a policy goal by the European Commission (CEC, 2004).

By the early 1990s, the OECD had also begun to explore the relevance of competitiveness to regions, through its emphasis upon the notion of structural competitiveness or innovation through the exploitation of science and technology in research (OECD, 1992; see also Cooke and Schienstock, 2000). By 1996, the OECD had included regions in its broad macroeconomic definition of the meaning of competitiveness, which it defined as 'supporting the ability of companies, industries, regions, nations or supranational regions to generate, while being and remaining exposed to international competition, relatively high factor income and employment levels' (OECD, 1996: p. 2).

Within the EU, it was successive British governments that made the running in promoting the discipline of competitiveness at both national and regional levels. This reflected the persistence of a productivity gap between the UK and its main competitors. In 1997, the Department of Trade and Industry (DTI) began to produce a bi-annual publication of 13 competitiveness indicators for the regions of England, plus Scotland, Wales and Northern Ireland classified into five categories relating to overall competitiveness, manufacturing performance, labour market outcomes, business development, and land and infrastructure (see Chapter 4). These were produced in response to demand for a wide range of regional indicators (and following consultation with a wide range of organizations), and evolved from the Business Competitiveness Indicators (BCIs) which were developed in 1995.

In 2003, the DTI in conjunction with the Economic and Social Research Council (ESRC) commissioned Michael Porter to investigate the current state of UK competitiveness. His report provided further impetus to the growing government emphasis on regional competitiveness: 'in the past, analysis of competitiveness has focused on the nation and national-level policies. Increasingly it is becoming clear that this perspective is too limited: competitiveness is affected by assets and policies at many different geographic levels. These range from cross-national e.g. the Baltic Rim of the EU, to national, to regional, to local. Indeed, the most significant spillovers and interactions take place at the regional and local level. A clear indication of the importance of regional business environments is the sharp performance difference across regions within given countries, even though they are all exposed to the same national level policies' (Porter and Ketels, 2003: p. 32). The report went on to say that 'in the US, for example, there is clear evidence that much of the relevant progress in improving the microeconomic foundations of competitiveness occurs at the regional level' (p. 32), and concluded that strong regional institutions were critical to delivering regional competitive success. This clearly chimed with dominant UK government thinking, since in 1998 it had established Regional Development Agencies (RDAs) for the English regions, which were explicitly tasked with responsibility for making their regions 'more competitive' and akin to benchmark competitive places such as Silicon Valley (House of Commons, 2000; HM Treasury, 2001).

Similar thinking was clearly emerging in Canada (ACOA, 1996) and the US (De Vol, 1999) at this time. In particular, 1986 saw the establishment of the US Council on Competitiveness, a forum for elevating national competitiveness to the forefront of the national consciousness and to lead the national policy debate on the subject. By the late 1990s, the Council was focusing on regional innovation and benchmarking and, with the notable influence once more of Michael Porter, launched the Clusters of Innovation Initiative 'with the realization that the real work of raising productivity and innovation capacity usually occurs not in our nation's capital, but in the cities and regions where firms are based and competition actually takes place' (Council on Competitiveness, 2001: p. v). To advance its innovation-based economic development model, the Council on Competitiveness subsequently launched the Regional Innovation Initiative (RII), which was explicitly designed to expand the Council's core messages regionally and reach more leaders in the private, public, university, labour and non-profit sectors.

The regional competitiveness discourse continues to spread further afield. For example, in 2004 a report for the Russian–European Centre for Economic Policy (which was funded by the EU) asserted that regional competitiveness had become one of the highest priorities on the agenda of the Russian Federation (Sepic, 2004). The lexicon of urban and provincial competitiveness has also recently begun to emerge in China, with the central government providing annual urban competitiveness reports since 2002 and a number of consultancies developing new benchmarking techniques (see, for example, Tan et al., 2008).

There are two dominant practical explanations for the rescaling of competitiveness from the national to the regional scale at this time, and thus the emergence of regions as defined spaces of competitiveness: the perceived importance of enrolling regions in national competitiveness agendas, and something of a crisis in the prevailing approach to regional policy and its emphasis on 'smokestack-chasing' as a mechanism for enhancing regional development and reducing disparities.

The EU was quick to identify the imperative of securing the support of regions in pursuit of its strategic competitiveness goals. Thus in 1999 the Commission observed that 'regional competitiveness is closely associated with four main factors: the structure of economic activity, the level of innovation, the degree of accessibility and the education level of the workforce. ... the association between these four factors and GDP per head across the Union suggests that if differences in their value between regions were eliminated, regional disparities in output would be reduced to around half their current level' (CEC, 1999a: p. 93). This was discussed more fully in the 2003 European Competitiveness Report which, in the context of EU enlargement to the east, provided a comprehensive review of the meaning of regional competitiveness and asserted the challenge of ensuring that 'the conditions that are necessary for regional productivity growth are developed across all regions including those of the acceding nations' (CEC, 2003: p. 129). The report went on to state that strengthening regional competitiveness throughout the EU was critical to the success of economic and monetary union and future economic growth, and was thus in the common interest.

In this regard, there has been a frequent emphasis within EU policy discourse on the links between competitiveness and the EU's other key objective of greater social cohesion. For example, the report of the second CAG emphasized the importance of competitiveness as helping to achieve the goal of social cohesion through the optimal use of each individual's potential, whilst EU regional and cohesion policy developed an explicit focus on investing in human and physical capital 'so as to raise the long-run growth potential of the weakest regions thereby improving competitiveness across the regions as a whole' (Hubner, 2005).

Similarly in the US, it was asserted that 'regional economies are the building blocks of US competitiveness' (Council on Competitiveness, 2001: p. v) and that 'the nation's ability to produce high-value products and services that support high wage jobs depends on the creation and strengthening of regional hubs of competitiveness and innovation' (p. 1). Likewise in Russia, the regional competitiveness imperative identified by Putin was couched in terms of the overarching need to double national GDP by 2010 (Sepic, 2004). In the UK a prevailing concern was the perceived threat to national economic objectives posed by evidence of widening regional disparities (Porter, 2003).

The growing dissatisfaction with traditional regional policy also helped to propel the growth of the regional competitiveness discourse inasmuch as there was a clear failure in policy approaches that competitiveness approaches seemed tailor made to address. Policies focused on holding down wages and reducing taxes, coupled with the provision of financial incentives to attract inward investment, were increasingly ineffectual in an ever more globalized economy with highly mobile capital and intensifying competition in respect of costs. They were also costly in terms of the level of expenditure required per job created (e.g. see Taylor and Wren, 1997). This encouraged talk of a 'transition' in developed economies such as the US and UK to a new policy phase focused on a process of 'upgrading' and 'redefining' the basis of regional policy from an emphasis on low input costs and an efficient business environment to an emphasis on indigenous capacities, productivity, knowledge and the development of more unique and innovative products and services (see, for example, Council on Competitiveness, 2001; CEC, 2003; and Porter, 2003). This quickly fed into policy. For example, in 2001 the UK government announced that it was introducing a new approach to regional policy which was intended to strengthen the capability of regions to build on their own competitive advantages 'by boosting regional capacity for innovation, enterprise and skills development' (DTI, 2001: p. 4).

The rescaling of competitiveness from the national to the regional level had two significant implications. Firstly, it led to the evolution of a polysemous regional competitiveness discourse. The transmission of a discourse originally applied at a national level and the perceived imperative of coercing regions into the support of national competitiveness objectives inevitably meant that regions were to a large extent considered to be mini-nations, subject to the same macroeconomic discourse and dominant conception of national competitiveness.

Yet at the same time, the evolution of regional policy in parallel with the emerging discourse of regional competitiveness also created a discourse that inevitably shifted away from the relative competitive positions of firms *within* regions, to the different resource endowments and degrees of attractiveness *between* regions and thus their ability to foster and attract innovative firms, skilled labour and investment. Competitiveness became something of a mix of the old notions of competing for mobile investment and the new Porterian notions of competitive advantage, productivity and innovation within the regional business environment (see Chapter 1). As a consequence, the discourse became one which did not imply a zero-sum game or win–lose competition, where the success of one region would happen to the detriment of another. In fact, the success of one region might create opportunities for other regions, especially neighbouring ones, owing to growth pole and spillover effects. In short, the belief that all regions could be winners in the competitiveness game became a widely held view, with regional competitiveness gradually emerging as a rather generic measure of 'relative success'.

Secondly, the rescaling of the competitiveness discourse meant that regions ceased to be perceived as passive players subject to the decisions of national or supranational governments, but were regarded as proactive players, enrolled in and in part responsible for delivering the competitiveness agenda and strategies for change. Regions became initially civic actors, enrolled in support of national competitiveness agendas. However, as the discourse was rescaled and evolved, regions became more aggressive and powerful strategic entities, motivated towards a common purpose and capable of influencing their own prosperity, performance and attractiveness. As such, the role of the state changed from one engaged in traditional industrial policy, towards the role of facilitator for developing innovative production systems at the regional level. At the same time, the emphasis on the attractiveness of the regional environment meant that regions also themselves became competitive entities, constituted and acted upon as self-disciplined actors needing to discipline themselves to the hypermobility of global capital (as Fougner, 2006; and Cerny, 2007). The regional state thus became commodified, entrepreneurial and saleable as a location for global investment, with regional actors increasingly engaged in a relentless and restless process of institutional restructuring in support of the competitiveness project (see Peck and Tickell, 2002).

Conclusions: the appeal of the competitiveness 'garbage can'

Using a cultural political economy approach, this chapter has examined the international origins and trajectory of the competitiveness discourse and its rescaling to regions. This has highlighted the rhetorical power and usefulness of the discourse for particular groups with particular interests, notably the business community, which has played a powerful role in promoting competitiveness agendas in policy. In this sense, there is clear resonance with the work of Hirschman (1986: p. 36), who states that interest-propelled action is characterized by two key features: self-centredness, meaning 'predominant attention of the actor to the consequences of any contemplated action for himself'; and rational calculation, that is, a systematic attempt at evaluating prospective costs, benefits, satisfactions and the like. Calculation is arguably the dominant element. Once action is informed by careful estimation of costs and benefits, with most weight necessarily given to those that are better known and quantifiable, it tends to become self-referential by virtue of the simple fact that each person or interest group is best informed about their own desires and satisfactions. Competitiveness is calculably in the self-interest of those wedded to neoliberalism.

The emergence of regions as a space of competitiveness also, however, reflects the power of the competitiveness idea as much as that of the interest groups promoting it. Thus, regional competitiveness has emerged as a highly elastic and yet dominant discourse because of the complex coming together of a range of different economic, institutional, pragmatic and political variables.

Indeed, the very appeal of the notion of competitiveness is precisely because it is abstract and lacks logical precision. As this chapter has demonstrated, the notion of regional competitiveness has acquired multiple meanings and definitions to fit the range of gaps in both national economic effort and regional policy that were identified as needing to be filled. In this context, the word 'competitiveness' has not needed a clear and precise meaning but has clearly served a useful galvanizing purpose for policy action, and (as Turner, 2001 noted) has provided a useful and recognized brand name from a previous era. In emerging as broad indicator of relative economic performance or success, regional competitiveness has thus become ambiguous enough to be flexible to suit the various demands being made of it. It has become a malleable policy garbage can into which a range of policy problems and solutions can be pitched.

Thus the power of the regional competitiveness discourse in part reflects its ability to present policy makers with a clear, coherent policy agenda which is particularly well suited to current manifestations of the regional state. Like globalization, competitiveness presents a relatively structured set of ideas, often in the form of implicit and sedimented assumptions, upon which they can draw in formulating strategy and, indeed, in legitimating strategy pursued for quite distinct ends (Hay and Rosamond, 2002). The regional competitiveness discourse points to a clear set of prescriptions for policy action within regions (detailed further in Chapter 3) and provides policy makers with the ability to point to the existence of seemingly secure paths to prosperity, as reinforced by the successes of exemplar regions.

The language of regional competitiveness also fits in very neatly with the ideological shift to the 'Third Way', popularized most notably by the New Labour government in the UK since 1997. This promotes the reconstruction of the state rather than its shrinkage (as under the purest form of neoliberalism) or expansion (as under traditional socialist systems of mass state intervention). Significantly, this philosophy sees state economic competencies as being restricted to the ability to intervene in line with perceived macro-economic or supply-side imperatives rather than active macro-economic, demand-side intervention – an agenda that is thus clearly in tune with the discourse around competitiveness.

All of this suggests that regional competitiveness is more than simply the linguistic expression of powerful exogenous interests. It has also become rhetoric. In other words, competitiveness is deployed in a strategic and persuasive way, often in conjunction with other discourses (notably globalization) to legitimate specific policy initiatives and courses of action. The rhetoric of regional competitiveness serves a useful political purpose in that it is easier to justify change or the adoption of a particular course of policy action by reference to some external threat that makes change seem inevitable. It is much easier, for example, for politicians to argue for the removal of supply-side rigidities and flexible hire-and-fire workplace rules by suggesting that there is no alternative and that jobs would be lost anyway if productivity improvement were not achieved. Thus, 'the language of external competitiveness ... provides a rosy glow of shared endeavour and shared enemies which can unite captains of industry and representatives of the shop floor in the same big tent' (Turner, 2001: p. 40). In this sense it is a discourse which provides some shared sense of meaning and a means of legitimizing neoliberalism rather than a material focus on the actual improvement of economic welfare.

This does not imply that there are no alternatives to the competitiveness agenda and that global capitalism and neoliberalism remain impervious to challenge. Rather, the implication is that the various alternatives that do exist as yet have lacked the same coherent, unifying characteristics and rhetorical power to dislodge competitiveness from its hegemonic status. Competitiveness is currently *believed* to be the only feasible option for regions and will conceivably continue to be so until alternatives emerge which command the same widespread degree of perceived rationality and can thus propel interest-related action.

Notwithstanding this, the CPE approach usefully reveals that the competitiveness discourse is not a global construct with an unequivocal meaning which is transferred through scales in a linear and entirely uncontested fashion. The evidence here suggests that, in being rescaled to the regional level, the discourse of national competitiveness has been subtly changed. Thus, for example, in enrolling the support of regions and broader civil society to the competitiveness cause in the EU, the discourse developed a characteristic binary with social cohesion agendas. Similarly, in the UK the regional competitiveness discourse has evolved in close concertation with the evolution of regional governance. In short, as the discourse is rescaled, it appears to be subtly recontextualized to fit each particular economic, political and social context. This in turn begs the question as to whether this recontextualization process simply reinforces the competitiveness hegemony and allows it to replicate, spread and increase, or whether it creates scope for contradiction, dissonance and challenge. Answering this question demands further exploration of both the mechanisms which have been developed to help spread the competitiveness discourse, and the relative importance of factors such as the power of the regional state which may challenge it. This thus provides the focus for Part II of this book.

Part II
Regional competitiveness in policy and practice

3 Competitiveness and the 'one-size-fits all' regional policy consensus

Introduction

With Part I of this book having explored the evolution and meaning of the concept of regional competitiveness, Part II focuses on problematizing the predominance of competitiveness (in its myriad conceptions) in policy terms at the regional scale. This chapter begins the process by focusing on exploring the implications of the dominant pursuit of competitiveness for supranational and national policy approaches to regional development intervention and governance.

As Chapter 2 has demonstrated, the adoption and spread of the discourse of competitiveness has followed directly from contemporary efforts to govern the economy in accordance with a neoliberal rationale. This has, in turn, privileged a particular 'attractiveness' conception of competitiveness which constitutes states and their populations as entrepreneurial and competitive 'place sellers' in a globalized marketplace for economic influence, resources and investment. The CPE approach has provided a useful mechanism for understanding how the notion of competitiveness has spread from nations to regions and, in so doing, has highlighted its power as a coherent, if eminently elastic, unifying discourse. However, this approach has also raised questions about the extent to which competitiveness is recontextualized when transferred to regions, and how and in what ways this changes both the meaning and the power of the discourse. Answering these questions demands further interrogation of the ways and means by which the competitiveness discourse is spread to regions, and the particular implications of this for regional policy and practice.

The purpose of this chapter is thus to consider the role played by top-down technologies of governance such as benchmarking and key 'discursive sites' in developing a particular regional competitiveness policy agenda. The chapter will argue that whilst there are powerful forces for policy imitation and thus, as a result, clear evidence of global convergence in regional policy discourse, approaches and institutional forms, the discourse of regional competitiveness contains certain paradoxes which suggest that the durability and extent of this policy convergence may have been somewhat overstated. The chapter begins by exploring how and why the discourse of competitiveness pro-motes convergence around a relatively narrow set of regional policy tools and approaches.

Benchmarking, imitation and policy transfer

The conception of competitiveness as the route to regional economic success, coupled with the transmission of its principal ideas by particular players and through key networks, is critical to understanding its potential to promote sameness in regional ideas and approaches. Since competitiveness is a relative concept, it implies the need to compare with others, such that regions are inexorably sucked into the continual monitoring and periodic benchmarking of what 'the competition' is doing and where the 'best offer' or success story lies (Malecki, 2004). Moreover, in being forced to compete for regional, national and EU funding, of which the rules of engagement prioritize the pursuit of innovation and competitiveness, practitioners have little option but to scour the horizon for what works elsewhere, in the hope of gaining favour from funding bodies (Haughton and Naylor, 2008).

National governments, the EU and international organizations have played a powerful role in facilitating this process and actively encouraging regions to benchmark themselves against other regions and to draw best-practice lessons through a range of tools including scoreboards, comparative indices of com-petitive performance and case studies of successful regions. The EU propa-gates this form of regional benchmarking as a new 'Open Method of Co-ordination' of regional policies. The Open Method of Co-ordination (OMC) was launched by the Lisbon European Council of March 2000. Under this system, EU member states jointly define common objectives but deliver them in the manner deemed most appropriate at national level. It is thus ostensibly a supranational strategy for engaging national governments in reforms inten-ded to promote competitiveness both for them and across the EU as a whole (Cammack, 2006).

The logic behind the OMC is simple. In policy areas where the treaty base for EU policy is thin or non-existent or where diverging political views hinder the development of law, modes of governance based on Council's guidelines, the coordination of national action plans, peer review of reforms, systematic benchmarking, performance indicators and governance processes open to the regional–local level and to civil society will, it is argued, result in European-wide convergence in incomes and living standards. This is otherwise referred to as 'deep Europeanization', which refers to a process of convergence with the growth and jobs objective of the Lisbon strategy for competitiveness without the need for new EU legislation (Getimis, 2003; Radaelli, 2008). The under-lying assumption is that open coordination encourages learning processes through discussion of indicators, national plans and peer-review processes of argument and persuasion which lead to the refinement of guidelines, timetables and goals. These learning processes are supported by a range of policy instruments which include fixed guidelines for EU policy with specific time-tables for achieving strategic goals; the establishment of quantitative and qualitative indicators and benchmarks against the best in the world, tailored to the needs of different member states and sectors as a means of comparing best practice; the translation of European guidelines into national and regional policies by setting specific targets; and periodic monitoring, evaluation and peer review or mutual learning processes (see Radaelli, 2008).

In 2004, the EU launched a major mid-term review of the Lisbon Agenda's progress with the inquiry led by the former Dutch Prime Minister Wim Kok. The Kok report focused on the EU's slow progress towards the 2010 Lisbon goals and pointed to particular disillusionment with the OMC, which it regarded as having 'fallen short of expectations' (Kok Report, 2004: p. 42). The report blamed member states for not doing enough to enforce bench-marking and made three key recommendations. Firstly, the EU needed to provide financial incentives to encourage member states to implement the Lisbon agenda. Secondly, partnerships for growth and employment should be established to help member states secure greater ownership of the Lisbon process. And thirdly, the report called on each member state to adopt a strategic approach involving a national action plan setting out its actions in respect of the Lisbon targets.

In the resulting renewed Lisbon strategy and more streamlined approach to OMC, regional policy was thus given a much more prominent role reflecting its potential to contribute to all three of these issues. Firstly, the Structural and Cohesion Funds were acknowledged to be the EU's main instrument for raising the long-run growth potential of the weakest regions and thereby improving the competitiveness across regions as a whole. In addition, embedded in the management system of European regional policy is the concept of partnership governance, which facilitates ownership of strategies by those responsible for designing them. European regional policy is also based on a strategic approach which, through multi-annual programming, obliges regions to think long term and identify strategic development priorities. Finally, in order to ensure the effectiveness of expenditure, the implementation of programmes includes specific management, monitoring and evaluation requirements. As such, EU regional policy became de facto a key tool in the Lisbon growth and jobs agenda, with the Commission ensuring that the EU15 member states were committed to earmarking at least 60 per cent of Structural Funds investment in convergence regions to growth and jobs objectives, and 75 per cent in regions under the Regional Competitiveness and Employment objective in the 2007–13 programming period.

Not surprisingly, these developments helped to spawn the growth of regional competitiveness indicators by which policy makers and analysts can measure, analyse and compare regional performance, or find out who is contributing to the Lisbon goals and effectively 'winning' the competitiveness game. (These are discussed in more detail in Chapter 4.) Monitoring, in particular, enables EU institutions to keep track of progress and compare what has been achieved by the 27 member states and their constituent regions both individually and as a whole. More specifically, with regard to regional policy, the EU has prioritized the pursuit of innovation which it sees as 'key to citizens' material concerns about their future' (CEC, 2006: p. 2). Thus, the EU has developed a range of mechanisms by which to facilitate policy learning in respect of regional innovation, including the European scoreboard, the European trend-chart, policy surveys, review and workshops on transnational policy learning, albeit with a light-touch approach (Radaelli, 2008). This reflects and reinforces the Commission's belief in the region's role in delivering broader competitiveness objectives, and in particular its assertion that 'the main competence to foster innovation often lies at regional level' (CEC, 2006: p. 16). As a consequence, EU regional policy has prioritized the pursuit of innovation, with regions encouraged to develop their own regional innovation strategies as well as be involved in the preparation of national reform programmes in support of the Lisbon competitiveness goals.

Startlingly similar technologies of governance have been deployed by the US Council of Competitiveness in pursuit of its competitiveness goals, with analogous outcomes ensuing for regional policy priorities. Following on from the publication in 2001 of the Porter-authored report on US regional competitiveness referred to in Chapter 2, the Council on Competitiveness asserted its firm belief in both the importance of innovation as the spur to competitiveness, and more especially the role of the region as the critical nexus for innovation-based economic growth. Regions were regarded in Porterian terms as the building blocks of national innovation capacity because of their ability to offer proximity and to facilitate the provision of the specialist assets critical to firm-level differentiation. As a consequence, the Council established the Regional Innovation Initiative (RII), which was designed to '[catalyse] consensus on policy priorities and practices to strengthen the regional platform for innovation' and to provide the 'tools and techniques that allow states and regions to inventory, evaluate and benchmark their innovation capacity' (Council on Competitiveness, 2005: p. 10). In support of this objective, the Council published a Regional Innovation Assessment Guidebook to help regional leaders assess the strengths and weaknesses of their regional innovation ecosystem, with a view to enabling them to understand, and then act to influence, the potential drivers of future innovation-based regional growth (Council on Competitiveness, 2005). This outlined a process for collecting data on key input and output metrics of innovation which could then be used to drive regional economic development policies and programmes. In addition, the guidebook metrics were designed to be used to develop an ongoing evaluation tool for regions tracking their innovation economy.

The methods described in this guidebook were tested and refined in six regional projects that formed the Council's Regional Competitiveness Initiative in 2003–5. The six regions that participated in the government-funded initiative were Central New Mexico; Northeast Ohio; Wilmington, Delaware; West Michigan; the Inland Northwest; and Greater St Louis. The Council claimed that through the foundational regional innovation project, the Clusters of Innovation Initiative, and collaborative efforts with the US Economic Development Administration (EDA), it not only helped to spur federal, state and local governments to enact pro-innovation policies, but also inspired private-sector leaders to implement innovation-based economic development strategies. Similar developments have taken place in Australia, where, since 1998, a consultancy firm – National Economics – has produced an annual stock-take of the economic well-being of Australia's regions and their prospects for economic development and employment growth. This is the so-called State of the Regions report (e.g. National Economics, 2008). The report encourages strategic emphasis within regions on regional competitiveness and highlights the scope and importance of intervention by regional government and agencies. Regular best-practice reviews and regional benchmarking exercises are also periodically undertaken or commissioned by the UK government, with one of the most recent being a comprehensive review of the competitiveness of Northern Ireland (Varney, 2008).

As well as benchmarking, the spread of the competitiveness discourse has been propelled by processes of policy imitation and transfer which have become an emblematic tendency in the field of regional development. Borrowing successful policies from elsewhere is regarded as a relatively quick and inexpensive way of fostering regional competitiveness. Not surprisingly, policy makers have inevitably focused their attention on a relatively small number of prototypical regional success stories, such as Emilia-Romagna in Italy and Silicon Valley in California (CEC, 2001; 2005), and a proliferation of so-called 'best practice' regional development research reports and papers have emerged as a result (see, for example, Institute of Welsh Affairs, 2001; One North East, 2006a). Particularly prominent have been attempts to replicate framework conditions for the formation of successful high-tech clusters and knowledge-intensive innovation systems, with many regions across Europe actively pursuing strategies designed to create the next 'Silicon Somewhere', such as Silicon Glen in Scotland, Language Valley in Belgium and Dommel Valley in Holland (Hospers, 2006) (considered in more detail below).

However, the fuzziness which surrounds both the concept of competitiveness and the relationship between key elements such as business clustering and competitive success has promoted the transfer of a rather generic and unspecific set of policy lessons. The discourse of regional competitiveness itself lacks clarity as to the precise significance of different determinants of competitiveness (as discussed in Chapter 1). As Doel and Hubbard (2002) demonstrate, the metaphoric use of a fuzzy concept can very easily produce a policy drift into the obsessive compulsion of atomistic regional analyses. Indeed, the competitiveness concept and theories on how to achieve it have been easily formulated into a range of stylized facts or black boxes which 'are far more portable and travel far more easily than the loose accumulations of empirical observations and theoretical inspirations that they represent' (Lagendijk and Cornford, 2000: p. 214). As a consequence, what has emerged is what Haughton and Naylor (2008: p. 170) characterize as 'pseudoagnosticism and highly selected empiricism' or an 'unholy alliance between "fast" policy transfer and rapid evaluation and judgement and a rush to replace and reinvent'. It seems that popular notions and rough rules of thumb about what works have been the main propellants of competitiveness-based policies, rather than sound empirical evidence on the forces driving competitive performance (see, for example, Romijn and Albu, 2002).

Thus, whilst many of the underlying theories and concepts warn against the uncritical adoption of strategies operating in successful economies of the world, these nuances tend to be lost in the translation into policy guidance (Harrison, 2006). The competitiveness literature has instead tended to proffer a generalized 'checklist' of relevant determinants of competitiveness, most of which are endogenous to the region and reside in the institutional environment (Deas and Giordano, 2001). These include, *inter alia*, the need for institutional thickness, a strong, local entrepreneurial culture, the availability of specialized suppliers, robust forms of inter-firm and extra-firm relations (e. g. with universities and research institutes) and the quality of the social, living and cultural environment – all things over which regional policy makers feel they exercise some degree of control.

This knowledge about the key determinants of regional competitiveness has, in turn, been codified into a relatively standardized recipe for regional success through being exchanged, shared and spread through the close networks of information exchange or 'reciprocal networks of interdependency' (Jessop, 2004) that characterize the regional development industry. Prominent organizations such as the OECD and the World Bank have played a powerful role in sedimenting the competitiveness orthodoxy, whilst academic consultants have increasingly been hired to offer relevant research and advice on appropriate benchmark regions and strategies. Governments have facilitated this process by bringing in overseas advisers and 'gurus' to speak at conferences and establishing funding for networks of shared ideas, inter-regional and international cooperation, and policy-learning processes (as, for example, in the EU's Interreg IIIC programme; see, Adams and Harris, 2005). There is thus a strong emphasis in regional policy on activating learning processes via organizational networks in a hierarchical and somewhat top-down form (Radaelli, 2008). This is typically then (re)circulated through the regional development industry or 'service class' (of development agencies, technology transfer centres, training organizations, consultancies and research institutes) via a plethora of conferences, workshops, seminars and symposiums and through their resulting papers and reports (Lagendijk and Cornford, 2000; Lovering, 2003; Boland, 2007). The same names and ideas dominate the policy context, so it is not surprising that policy imitation flourishes. In essence, a global market in commodified policy ideas has emerged, grounded in a shared cognitive and normative discourse of competitiveness, which is circulated through networks of consultants, policy makers and academics.

The 'one size fits all' approach: cloning of strategies and institutional arrangements

In practice, the processes of benchmarking and policy transfer lead inexorably to the establishment of regional policies with similar objectives, policy concepts and instruments (Hospers, 2006). The tendency for replication in development agendas is further reinforced by the need to be attuned to market imperatives. Regions are inevitably pushed towards homogenization of the place product because the market in which they are competing is the same (globalized) set of firms, investors, tourists and consumers (Smith, 2001; Malecki, 2004). This, coupled with broader tendencies for institutional transplantation (De Jong et al., 2002) and lesson drawing (Rose, 1993), provides a strong impetus towards convergence on neoliberal and essentially supply-side forms of policy approach and governance (Fougner, 2006). As a consequence, it is increasingly understood that regional development policy exhibits a strong tendency towards convergence and isomorphism, or what Malecki (2004: p. 112) terms 'the serial reproduction, the imitation and replication of the same ideas from place to place'. The result is a 'one size fits all' approach to regional economic development policy and a developing 'ubiquitification' of strategic approaches (Maskell and Malmberg, 1999). Such approaches are also increasingly being rescaled to apply to cities and the emergent 'city-region', where the same amalgam of fashionable ideas is being imitated from place to place (Harrison, 2007). On the surface at least, it would appear as if regions, cities and, increasingly, city-regions in Europe, North America and much of the rest of the developed world are pursuing 'identikit' development strategies.

Convergence is, however, a complex concept and there are different degrees to which it can occur in policy fields (Pollitt, 2001). The simplest level at which it can occur is that of discourse with the emergence of a shared language. Beyond this, there is cognitive convergence, which refers to the identification of a common set of beliefs about the key policy problems and the causal mechanisms and relationships at work in a particular policy area. This begs the question of how deep is the convergence process that is at work in regional development policy and practice.

There is certainly widespread evidence of global convergence in regional policy discourse and strategy objectives (as already demonstrated in Chapter 2), as well as cognitive convergence around the broad grouping of policy tools and institutional forms essential to improving regional competitiveness (as will be demonstrated here). Indeed, in an overview of international trends in regional policy, the OECD (2005) asserts that regional policy has evolved from a top-down, subsidy-based group of interventions designed to reduce regional disparities, into a much broader 'family' of policies designed to improve the competitiveness of regions.

In particular, there is clear agreement around the importance of two key elements to this policy approach. The first is the imperative of having a development strategy outlining the competitiveness goals for a region, and identifying the wide range of direct and indirect factors deemed to impact upon the performance of local businesses. There is a clear belief that 'successful regions build strategies' (Council on Competitiveness, 2001; p. vii) and a strong top-down impetus through policy funding and programme requirements, particularly in the EU, to foment regional plans and strategies.

The second is the importance of a collective or negotiated governance approach led by regional and local government and key stakeholders such as regional development agencies, with national/central governments playing a less dominant role. According to Harrison (2006: p. 33) regional development strategies are characterized by the 'cloning of an amalgam of institutional arrangements coveted from successful regional economies', whilst Lagendijk and Cornford (2000) refer to the highly isomorphic organizational field that has emerged in regions internationally because of the market through which organizations compete, the professional networks on which they draw and the mimicking of peers, new models and ideas. As a consequence, regions are increasingly dominated by supply-side, partnership-orientated institutions and networks of associative or collaborative governance (see Box 3.1 for some illustrative, international examples and analyses).

Box 3.1: Regional competitiveness strategies: some international examples

The US experience: The US Council on Competitiveness has played a hugely influential role in propelling the development of regional competitiveness strategies based around innovation and institutional collaboration across US states and metropolitan regions. As well as its range of competitiveness-orientated analyses and reports, the Council has developed a very close working relationship with the federal government's Economic Development Administration (EDA). Such is the dominance of the regional competitiveness discourse here that the EDA's stated agenda is to promote innovation and competitiveness and prepare American regions for growth and success in the worldwide economy. Put simply, regional competitiveness has become the ultimate strategic goal.

In acting upon this, between 2003 and 2005 the EDA sponsored Regional Competitiveness Initiatives, in conjunction with the Council, across six regions: North East Ohio; Central New Mexico; Greater St Louis; Wilmington DE; Inland North West; and West Michigan. This initiative was designed to assist each region with an inventory and evaluation of its innovation assets, strengths and weaknesses; build a consensus amongst regional leaders on priorities, policies and practices for strengthening the regional innovation platform for competitiveness; and to host regional competitiveness summits to build regional support for action on the key issues identified in the regional analyses. The West

Michigan evaluation, for example, highlighted the need to improve and deepen the nature and extent of collaboration between the region's firms, governments and educational institutions. This subsequently led to the establishment of the West Michigan Regional Competitiveness Initiative and the institution of three Regional Action Teams to implement the resulting development strategy. This in turn spawned the development of the West Michigan Economic Development Partnership of the region's economic development professionals committed to collaboration and cooperation across the region. This was an effort to emulate the experience of a range of successful joint public–private coordination, planning and policy-promotion agencies that have worked elsewhere in the US, such as the Chicago Metropolis 2020 Plan and the Silicon Valley Manufacturing Group.

In 2007, the EDA provided further funding for the Council on Competitiveness through its support for a new regional leadership initiative designed to boost innovation-driven regional development and empower collaborative leadership networks to create a competitive advantage for American companies and workers (see www.compete.org). Further evidence of the convergence of key ideas and discourse is evident by the establishment in September 2006 of the Centre for Regional Competitiveness in the University of Missouri, which has a stated aim of helping regions pull together the pieces of a winning strategy and build the strong regional partnerships deemed essential to leverage the resources and skills critical to competitiveness (see www.rupri.org/regionalcomp.php).

The Canadian experience: British Columbia: Markey et al. (2008) have undertaken a critical review of 15 community and regional economic development strategy reports in north western British Columbia, Canada between 2005 and 2006. This has provided an interesting insight into the degree to which the competitiveness discourse has pervaded strategic thinking in what is a relatively rural, natural and resource-rich region of Canada. The authors find that economic strategy reports in the region provide 'a multitude of generic recommendations for the region associated with how to promote competitiveness' (p. 342). The concept of competitiveness is, however, both poorly defined in these strategies and poorly understood by those charged with designing and delivering the strategies on the ground. The economic strategies employed tend to 'borrow an understanding of local development and competitive variables from other places and sources. In terms of competitiveness, this means that a relatively standard list of variables is mentioned – variables that are often more closely associated with urbanized environments' (p. 343).

The Australian approach: Collits (2004) has demonstrated the convergence of Australian regional policy with the international policy norm. Since the abandonment of a national regional policy characterized by efforts to counter the dominance of key metropolitan regions, Australian regional policy approaches have converged to something of a consensus since the 1990s. The key feature of this approach is its emphasis upon locally driven strategies and objectives working in partnership with national regional policy. It also acknowledges the limitations of national/ federal government policy and emphasizes the role of regions and their communities in committing their own resources to the improvement of their respective regional economies. The consensus is characterized by a belief in regional competitive advantage as the way forward for regions; the development of partnerships with and among regional and local development agencies; tailored, region-specific policy approaches; competitive grant funding delivered to a plethora of 'meso level' agencies operating at the regional level; a strong preference for growing existing regional firms rather than recruiting new businesses from outside the region; and a belief that managing change is the key regional policy objective. This belief is clearly evident in New South Wales, for example, where the 2006 State Plan emphasizes the importance of developing a small business community 'that competes with the best in the world and wins' (New South Wales Government, 2006: p. 88) and which prioritizes improved innovation and enhanced productivity.

The UK approach – Example 1: A winning Wales: In 1997, Wales saw the establishment of its own devolved regional government. Keen to develop policies which were made in Wales and for Wales, the Welsh Assembly Government published a 10-year economic development strategy for the region in 2002, with the express aim of improving its competitiveness and reducing the relative prosperity (GDP) gap between it and the rest of the UK (Welsh Assembly Government, 2002). 'A Winning Wales' clearly resounded with the discourse of competitiveness and set demanding targets for the Welsh economy, including improving enterprise and innovation, raising skill levels and learning performance, and using world-class electronic communications to the maximum potential. The strategy also emphasized the development of a better coordinated and well-targeted business support network. The Welsh Assembly Government took responsibility for delivery of the strategy in conjunction with the Welsh Development Agency and Education and Learning Wales, with these bodies ultimately absorbed by the Assembly in a so-called 'bonfire of the quangos' from 2006. At the same time, 'A Winning Wales' was replaced by 'WAVE: Wales a Vibrant Economy' (Welsh Assembly Government, 2006) (see also Chapter 5).

The UK approach – Example 2: Leading the way: One North East: In the UK, the national government in 1999 explicitly tasked the newly established Regional Development Agencies for the English regions with responsibility for making their regions 'more competitive' and akin to benchmark competitive regions (see DTI, 1999). The Regional Economic Strategy for the North East of England (One North East, 2006b) provides a typical recent example of the competitiveness strategies that have emerged as a result. The focus of the strategy is on raising business competitiveness and closing the GDP gap with the rest of the UK, with once again an emphasis on encouraging business start-ups, raising business productivity and improving education and skill levels. This is designed to continue the North East's transition 'from a largely industrial to a more knowledge-based economy, where skills provide a competitive edge' (p. 10). Particular emphasis is placed upon the importance of 'collective regional leadership'. Indeed the strategy notes that 'central government can only help close the productivity and participation gap to an extent. We must develop stronger leadership and collective responsibility for improving our economic performance. Sectors and organizations must work collaboratively to challenge and champion North East England' (p. 4).

The German Standortdebatte: Brenner (2000) provides an insightful analysis of the emergence of the regional competitiveness discourse in Germany as part of the nation's broader adoption of the competitiveness (*standort*) discourse during the 1990s. This reflected the assertion that the German economy had become uncompetitive within the EU and that the mobilization of region-specific strategies and institutions was critical to strengthening the nation's competitive advantages as a whole. As such, the promotion of national competitiveness was seen to hinge upon 'the construction of "Euro-regions" associated with territorially specific conditions of production, socio-economic assets and institutional forms at sub-national scale' (p. 321). Thus, national state institutions acted as animators and mediators of processes of regionalized strategy and institution building, assisted by a range of neoliberal socio-political forces, including regional politicians and bureaucratic elites, major business organizations and a range of think-tanks and research institutes. Collectively, these bodies played a major role in redefining national state spatial planning policy as an instrument of competitiveness by privileging capitalist growth over other socio-political goals, aggressive promotion of inter-spatial competition, and the treatment of uneven regional development as a natural basis for capitalist expansion. Thus, each of Germany's major metropolitan agglomerations was encouraged to emphasize and develop its regional potentials and capacities. This was supported by the devolution of tasks and responsibilities from states to regions, and specific support and encouragement for major institutions and actors within metropolitan regions to coordinate their activities. In effect, Germany has witnessed the development of 'a "glocal" developmentist project, based on strategies to enhance global competitive advantages by splintering national economic space among highly specialized regional and local economies' (p. 332).

The cloning of policy instruments

Within this broad emphasis on strategy, there is also convergence across regions around the broad set of policy tools and interventions deemed critical to improving competitive advantage. Developing strategies to impact upon the competitiveness of a given region involves identifying the sources or potential sources of a region's competitive advantage. As such, an extremely wide range of factors could be targets for policy, some of which may be international or national in nature and lie beyond the scope of regional strategies, whilst others appear more amenable to influence at the regional level.

Broadly two sets of policy approach have become dominant in the head-long pursuit of regional competitiveness: policies and instruments to increase the competitiveness or productivity of firms in the region; and policies and instruments to improve the quality of the region as a business location (OECD, 2005). Each of these in turn is now explored in more detail.

Policies targeting firm-based innovation, productivity and competitiveness

Firstly, a ubiquitous feature of regional competitiveness strategies is those policies which provide direct support for enhancing the competitiveness of individual firms within regions. These work in a range of ways. Some interventions work by directly assisting businesses with a range of factors shaping their own performance, such as measures to support improvements in their internal organization, management styles and structures, product development, marketing and advertising and so on. Recent years have indeed witnessed a proliferation of business-support initiatives and agencies across regions, providing a range of specialist business advice and services for business start-up, growth and improvement, often in the form of regionalized network agencies or 'one-stop shops' such as Business Connect, or more recently, Business Link in the English regions, for example. This has, in turn, contributed to the growing array of institutional forms of economic governance in regions and the emergence of an increasingly complex and congested state (Skelcher, 2000) (see also Chapter 5).

Even more ubiquitous, however, has been the growth in policy initiatives emphasizing the innovation capacities and potential of groups of firms in the region, their more effective use of available knowledge and technologies, and their relationships with the business environment within which they operate. This reflects the evolution of the body of work around regional innovation which has placed particular emphasis on the importance of physical proximity between firms, and the development of a shared regional culture of particular norms, understandings and conventions which facilitate the exchange and sharing of proprietary forms of knowledge. This has paralleled developments in thinking around national innovation, such that national and regional innovation policies have progressively converged over time (Hassink, 1993).

Two particular sets of literature provide the basis for much of this policy development. The first of these relates to the body of evidence which emerged from the 1980s onwards as to a range of post-Fordist examples of successful flexible specialization in new industrial districts. Analyses of these 'new industrial districts' or spaces helped to propagate belief in the importance of various untraded, non-material interdependencies between firms to their competitive success. These are the collective advantages which stem from the historical development of local sectors and links with the region, firm size and structure, the level of specialization, the use of advanced technologies and networking and information exchange between firms and related organizations (Storper, 1995). The second seminal work was Porter's diamond model of firm clustering (see Chapter 1). This proved particularly appealing not only because it served to explain how and why firms cluster together, but also because it provided an eminently replicable and thus saleable framework for policy makers to use to diagnose the innovation strengths and weaknesses of firms, and their interrelationships with one another and the regional environment, as well as to identify pathways for their future development and growth.

In respect of their development of policy instruments to operationalize regional innovation through the construction of proximity and enhanced knowledge exchange and interaction, regional policy makers have overwhelmingly focused their efforts on three categories of policy measure (following OECD, 2005): real estate-based projects; relational asset or cluster policies; and policies to link research and business.

Real estate projects

Real estate projects, designed to create a shared industrial space for technology development, transfer and networking through co-location, have a relatively long history, having come to the fore in the 1970s and 1980s. A number of different approaches have emerged over time. Early interventions typically focused on the creation of large, publicly funded technopoles which were often designed to attract inward investment, such as the Research Triangle in North Carolina, US, and Sophia-Antipolis near Nice in France, home to a number of significant telecoms firms (OECD, 2005). Over the 1980s and 1990s, the science park model became prominent, with an emphasis upon supporting research and development capacities and exploiting technological creativity. These have typically been established on or near university campuses (such as the Cambridge Science Park, and Surrey Research Park in England). More recently, such approaches have emphasized business incubation and entrepreneurship through the construction of business complexes to house independent small and medium-sized enterprises (SMEs) (for example, the range of Techniums, established in Wales by the Welsh Assembly Government and Welsh Development Agency in 2001 to provide an environment within which new and early-stage science and technology businesses can flourish).

Cluster policies

The archetypal and arguably most popular and prominent of the policy measures for encouraging regional innovation are cluster policy initiatives. These are designed to support existing or nascent agglomerations of firms through the provision of collective services and other measures designed to build cooperation and encourage collaborative efforts with business activities such as marketing, exporting and so on. Indeed, the extensive reach and popularity of these policies is such that they have become almost 'axiomatic' (Palazuelos, 2005) and certainly a pre-eminent economic development approach in numerous regions across Europe and beyond (see, for example, DTI, 1998; OECD, 2001; CEC, 1999b).

For example, Japan has promulgated a national cluster policy following the success of locally driven clusters in Tokyo, whilst in Montreal in Canada, policy makers have sought to establish a more comprehensive cluster development strategy to develop new clusters beyond existing strengths in aeronautics and pharmaceuticals. In the UK, Scotland was one of the first regions to develop a cluster initiative, with cluster audits undertaken regularly by the development agency, Scottish Enterprise, from 1993 onwards (OECD, 2005). This has since been replicated across many other UK regions, with, for example, the One North East Development Agency in North East England instituting a cluster development action plan in line with its belief that cluster development represents 'a common sense approach' to achieving its regional competitiveness and growth objectives (One North East, 2007).

The proliferation of cluster policies reflects, in part at least, the perception that cluster policies (unlike real estate projects) are a relatively low-cost regional development and enterprise policy (OECD, 2005). It also, however, reflects the widespread benefits that are expected to ensue from successful clusters. Porter (1998: p. 197) defined clusters as 'geographical concentrations of interconnected companies, specialized suppliers, service providers, firms in related industries, and associated institutions (e.g. universities, standards agencies and trade associations) that compete but also co-operate'. Thus a substantial array of benefits over and above standard agglomeration economies are deemed to accrue to firms located within clusters. Specifically, clusters are deemed to result in higher productivity for firms within them and to encourage the cross-pollination of ideas and innovation between firms in the cluster, often through high rates of inter-firm labour mobility. Moreover, the cluster itself is widely believed to have the added potential to generate knowledge spillovers, collective learning and higher rates of new firm foundation, of benefit to the wider regional economy. As such, it is not surprising that cluster policies have proved to be a highly attractive and seemingly 'magic bullet' policy option for policy makers seeking to enhance innovation and regional competitiveness (Burfitt and MacNeill, 2008).

Cluster policy interventions vary significantly by industry, sector, innovation model, cluster life stage and the degree of networking they embrace. Policy manifestations range from policies designed to encourage low-resourced, small-scale business networks without a specific sectoral focus, to more complex, large-scale programmes of coordinated measures designed to target a specific, geographically cohesive industry. As an approach to regional development, cluster interventions differ from traditional incentive-based regional development policies inasmuch as they concentrate their support on networks of diverse agents rather than on individual firms. As such, they are regarded as being innovative in respect of their holistic approach (see OECD, 2005).

Policies to link research and business

Policies explicitly seeking to improve the interrelationships between business and research institutions, especially universities, have also proliferated in regions. Until recently, universities were typically viewed as essentially being providers of basic knowledge for the labour market, or 'ivory towers' with few, often only arm's-length, connections to their immediate regional environment. Universities are, however, increasingly acknowledged as playing a multi-faceted role within local and regional economies. As well as acting as key sites of tertiary learning and research, they are also typically major employers and purchasers of services, as well as providers of cultural, recreational and infrastructural resources. They also act as an integral part of the city or regional network of public facilities that act as centres of attraction for individuals and enterprises (Lambert, 2003).

More pertinently, the more indirect knowledge-creation impacts of universities in local and regional development have assumed an elevated status with the emphasis on regional innovation, knowledge and competitiveness. This reflects the perceived importance of their role in a number of successful high-tech clusters, through activities such as technology transfer, licensing agreements, technology parks, the development of a labour pool of talented scientists and graduates, and the creation of spin-off firms (see, for example, Mayer, 2006). As a consequence, policy makers have increasingly encouraged higher education institutions to develop and strengthen their links with business and industry through a range of initiatives and new venture partnerships, with a particular emphasis on encouraging the commercialization of research.

University–business collaboration is particularly important in the US, where there is a strong tradition of well-organized technology transfer and entrepreneurship incentives throughout the university sector, and strong partnerships with private and other public organizations. Similar patterns are now also emerging in many other countries, notably across Nordic Europe, often linking with elements of the technology centre and cluster policy approaches outlined above (OECD, 2005).

Policies targeting the quality of the regional business environment

Alongside the increasing recognition that regions are not just the passive objects of the location decisions of firms but are also constitutive of actors who act in their own interest by trying to keep or attract firms (see Chapter 2), a second broad set of policy interventions has emerged which focus on enhancing regional assets and the local 'milieu'. This refers to the regional business environment which facilitates both clustering and an innovation-orientated and risk-taking business culture. These policies are thus complementary to, and provide the necessary support for, cluster and other innovation-based interventions.

The wider business environment at regional level includes a range of factors that enable business activity and which can thus be influenced by strategic policy intervention. These include so-called 'hard' factors such as the efficiency of the transport and communications infrastructure and the local tax regimes, which may impact upon the costs of conducting business in the region. Increasingly, however, these policies are focusing on place-specific externalities or 'soft' location factors that have the potential to create a favourable environment for the development of local firms and thus to improve the accessibility and attractiveness of the region as a whole – or to make places 'sticky' for mobile capital and skilled workers (Markusen, 1996). These include interventions in the realms of environmental quality, social stability, cultural resources, educational institutions, the quality of the living environment, and the regional identity and international image (Cappellin, 1998) – all of which are attributes derived from good local governance. This, in turn, tends to encourage regions to invest in upgrading those natural and man-made amenities which differentiate them and which provide the foundations for a range of economic activities, such as local food production and tourism (OECD, 2005).

Regions everywhere are increasingly engaged in a variety of 'place-making' improvements in the built and natural environment. Such efforts to improve business premises, housing, key leisure, cultural and retail facilities and the public realm are clearly actively promoted by governments, think-tanks and development agencies. For example, the US Council on Competitiveness (2006) talks of regional competitive success as ever more dependent 'on the quality of ideas and talent' (p. 7). Likewise, the place-making theme is a core element of the UK government's recent sub-national review of economic development (BERR, 2008a), which focuses on enhancing the 'liveability' of places through improvements in the physical and social environment. This approach has for some time dominated urban development agendas, where redevelopment projects and regeneration programmes have become the norm to enhance the city image via flagship developments. In particular, recent years have witnessed a growing emphasis on cultural projects, events and attractions following the popularity and influence of Richard Florida's work on the economic benefits to be reaped from attracting the most talented workers or 'creative classes' (Florida, 2003). Perhaps not surprisingly, as a result, the urban development literature is increasingly punctuated with concerns that cities, like regions, are lapsing into copycat strategies and becoming awash with the same flagship sport and leisure complexes, often based on similar designs – what might be regarded as 'karaoke architecture' (Evans, 2003).

The paradoxes of benchmarking and uniqueness

The undeniably powerful forces for convergence in regional policy and practice around the discourse of competitiveness must, however, be understood in relation to the countervailing forces for policy variation. Indeed, the competitiveness discourse is, arguably, riven with paradoxes and contradictions that have the potential at least to undermine the pressures for ubiquity and sameness.

On the one hand, benchmarking practices carry a number of contradictions. Firstly, policy interdependency and the need to coordinate strategy efforts across regions in support of national or supranational policy goals (such as the Lisbon agenda) create tensions between the desire to encourage regions to develop at their own pace (with greater diversity as the likely outcome), and the objective need to steer the process towards convergence on strategic policy goals. Similarly, open methods of coordination and networked approaches to policy transfer and learning tend to propel more cooperative forms of learning (where everyone pursues the same models and lessons) and thus convergence, whereas more competitive learning through encouraging regions to adapt lessons from exemplar regions themselves would be more likely to result in more divergent outcomes, i.e. enhanced competitive outcomes for some regions yet clearly not for others (see Radaelli, 2008).

Perhaps more pertinent is the key paradox which lies at the heart of the regional competitiveness discourse – what might usefully be referred to as the paradox of uniqueness. The paradox can be described as follows. Central to the theory and discourse of regional competitiveness is the assumed importance of so-called region and localization economies, or the factors within the regional business environment and milieu that shape the functioning and performance of the region's firms, their capacity to innovate, the scope for clustering and the potential for enduring and positive synergies to be cultivated between business, research establishments and institutions of governance. In practice, however, the region's influence may vary, depending on the particular industrial structure and context, the balance of globally and locally oriented firms, and the degree to which the region constitutes an internally cohesive, homogenous economic space. The critical point is that the nature and significance of this contingency is barely acknowledged in the regional competitiveness discourse. There is thus an inherent paradox here, in that whilst the discourse emphasizes the importance of factors endogenous to the region in shaping firm performance, the key ingredients for success are uniformly prescribed.

Furthermore, even though the quality of place is increasingly articulated as part of the growing emphasis on the social environment in the policy discourse, it is a notion of 'place' that elides place liveability with place marketability and thus has a clearly dominant economic rationale and end product in mind – that is, the attraction of investors and people (Jarvis, 2007). The result is a 'vicious circle of conflicting interests' (Harrison, 2006: p. 33), where regions are drawn to policy measures informed by the experiences of prosperous regions that reside firmly within their embedded local inter-dependencies. Not surprisingly, urban areas are suffering the same intrinsic contradictions whereby competition between them has led to a multiplicity of standardized attractions that reduce the uniqueness of their identities whilst claims of uniqueness grow ever more intense (Zukin, 1995).

The collective implication of these contradictions for regions is that if they try to emphasize their uniqueness in this context, they run the risk of severe problems of legitimacy and accountability, undermining their place in the competitiveness game. This raises a number of questions around the scope that exists, if any, for regions to develop policies and institutional arrangements that break significantly from the norm. Is it possible for regions to prioritize, tailor or even resist the dominant policy prescriptions of competitiveness to suit their specific regional contexts and circumstances?

Conclusions: the limits to convergence

This chapter has demonstrated that there is clearly strong evidence of inter-national convergence around the discourse of competitiveness in regional development policy, coupled with significant evidence of cognitive convergence around the causal mechanisms deemed to be at work in shaping regional competitiveness outcomes (in particular, belief in the central role played by firm and region-wide innovation). This, in turn, has created powerful forces for convergence around both the key policy levers for delivering competitiveness (regional strategies and a collective, region-led governance approach), and the specific policy tools and instruments to be deployed within them (policies to improve firm innovation and clustering, and interventions to improve the attractiveness of the regional business and living environment). This process has clearly been facilitated by processes of benchmarking, policy transfer and the compulsion to emulate 'best practice'.

The policy variables that have become ubiquitous have certain character-istics which make them relatively easy to codify and replicate. This is because they are either process oriented (concerned with discourse, strategy develop-ment and network construction) and/or relatively inexpensive and easy to implement (as in the case of cluster strategies and action plans). They are also elements that typically and readily lend themselves to mediation by local and regional governance structures and agencies. When these facts interplay with the evident pressures for coherence in policy goals and economic performance across different spatial scales, the power of the pressure for convergence is readily manifest.

This does not, however, mean that the process of convergence is completely universal in coverage or that regional development approaches are indeed everywhere the same. There are certainly parts of the world where policy approaches in regions differ from the competitiveness norm. For example, the extent to which the provinces of China are pursuing their own competitive-ness strategies or to which their plans are clearly and simply replicating those of regions in the western world, remains an open question. Clear differences in land and property ownership and markets necessarily create scope for difference in what competitiveness means and how it is deployed. Nevertheless, the Chinese provinces are beginning to develop their own economic identities and interests and, in the process, are clearly engaged in significant competition for preferential policies and for commercial investment. In the context of the expanding decision-making powers of the provinces, an emerging field of study is thus the analysis of the business strategies employed by provincial governments in pursuit of competitive advantages over other provinces (Hendrischke, 1999: p. 1).

More significantly, it cannot be assumed that regional development policy makers and practitioners wedded to the same competitiveness discourse will necessarily implement policies in the same way or with the same outcomes. As Radaelli (2008: p. 249) observes, 'convergence in "talk" may not produce convergence in decisions'. Other variables may clearly come into play and shape both the transmission of policy ideas into tangible interventions and models of economic governance, and the ultimate nature and effectiveness of these interventions in practice. Indeed, there may be something of an 'implementation gap' between regional competitiveness strategies and action plans and collaboration efforts on the ground, especially where such strategies are not sufficiently grounded in the reality of the local context (Markey et al., 2008), or where regional institutions lack the required leadership skills and organizational capacity to effect them adequately (Gibbs et al., 2001). Furthermore, there is growing evidence to suggest that the various competitiveness-orientated policies outlined here do not present the magic bullet that policy makers may require or indeed believe in. Many technopoles, for example, have been criticized as little more than expensive 'cathedrals in the desert', whereas the ingredients that make for successful clusters often lie within business-orientated action which the public sector finds difficult to initiate or recreate. This points to a series of potential flaws in competitiveness strategies and policies and, in turn, raises serious questions around the suitability of regional competitiveness and its seemingly unending hegemony.

4 Performance indicators and rankings

Deconstructing competitiveness league tables

Introduction

As the previous chapter indicated, the predilection for benchmarking and the spread of the competitiveness obsession have created a voracious demand for indicators by which policy makers and analysts can measure and compare regional competitive performance. Quite simply, there is a desire to see which regions are 'competitive', or 'winning' the competitiveness game. This raises all sorts of questions around what it means for a region to be competitive and how this can be measured. Efforts have increasingly focused on the development of composite indices which bundle together into one overarching measure all those indicators deemed to be relevant to competitiveness. These have the clear advantage of presenting results which can be ranked and thus reported in the form of a 'league table' – thus providing a very visual means of portraying regions as competitive entities, competing with one another for the prize of competitive success. These are growing in popularity, and indeed a plethora of city and regional indices are now in evidence around the world on the basis of a range of different measures of competitiveness. Perhaps not surprisingly, such indices and rankings attract widespread attention in the media and are increasingly used and indeed funded by governments as part of their general analysis of the performance of regional economies and their key institutions.

However, significant questions surround the validity and usefulness of these indices and rankings and what they actually tell us about regions, their competitive performance or success and what this 'success' actually means – questions which to date have received only limited critical interrogation.

Some of these questions relate to the practical task of whether and how regional competitiveness itself can be measured. The chaotic and fuzzy nature of the competitiveness concept and the theory underlying it suggests that this is unlikely to be a straightforward or easy task, with critical questions, among others, concerning what variables should be included in the indices and for what 'regions', and how (if at all) these indicators should be weighted and aggregated to produce an overall index.

Beyond this, even if regional competitiveness were to be properly measurable, it is debatable what any measures of regional competitiveness actually tell us that is meaningful. There are, for instance, considerable questions around the potential utility of these indicators as a means of helping firms, policy makers and institutions to assess the performance of their economies in comparable (i.e. numerical) terms, and to undertake appropriate remedial strategies in response. In a similar vein, there are questions as to whether league tables – which, by design, imply directly comparable performance between fixed or known entities competing directly in the same contest for the same prize – are indeed meaningful for the complex, putative open and multifaceted entities that constitute 'regions' (see Introduction). More fundamentally, there are considerable questions around what these indicators and league tables reveal that is meaningful about a region's 'success', who benefits from them, and what this means more broadly for the quality of life and wellbeing of those living and working in regions. To put it starkly, what is this 'competitiveness' for?

These are some of the principal questions to be explored in this chapter, which now proceeds as follows. It begins by examining the growth in the number and range of indices of regional competitiveness, with a view to understanding their function and appeal. It then draws on some of the most frequently reported composite indices of regional competitiveness to explore in more detail how they are typically constructed, the conceptualizations of competitiveness on which they are based, and their utility for policy makers and analysts. It then proceeds to provide a critique of the adequacy of competitiveness indices in providing meaningful information about regional 'success' and development when these terms are conceived of more broadly.

The growth of competitiveness indices

The pursuit of competitiveness as a key policy goal has created enormous global interest in the measurement of competitiveness at all spatial scales. Indeed, there is a huge empirical literature around different likely indicators of an economy's competitive performance or success, some of which focus on 'revealed' performance measures such as GDP per head, productivity, terms of trade, relative unit labour costs and so on, and others of which seek to construct and operationalize more sophisticated and specific measures of 'competitiveness' itself.

The measurement of national competitiveness is a relatively well-established phenomenon, with a broad distinction evident between analyses which consist of the reporting of a series of separate indicators, and those which seek to develop composite measures and rankings. Among the ones best known to business leaders and policy makers are those prepared annually by the International Institute for Management Development in Lausanne (the IMD), and the World Economic Forum (WEF). Since 1989 the IMD has measured national competitiveness with regard to countries' performance in providing an environment that sustains the domestic and global competitiveness of the firms operating within their borders. The report seeks to form a view of the

competitiveness of nations by examining a multiplicity of factors, both quantitative and qualitative (the Competitiveness Scoreboard), and obtaining the
views and perceptions of business people themselves about how their own
nation performs relative to other nations (Executive Opinion Survey). The
approach used distinguishes four main competitiveness factors: economic performance, government efficiency, business efficiency and infrastructure. The
analysis focuses on the competitiveness of 55 nations (and 9 regions) based on
some 331 competitiveness criteria (both quantitative and qualitative data).[1]

Similarly, the WEF has been publishing its World Competitiveness Report
annually since 1979. The objective of this report is to assess the comparative
strengths and weaknesses of national competitiveness in terms of competitiveness and prospects for growth. It is constitutive of three index rankings –
the Growth Competitiveness Index (GCI), the Business Competitiveness
Index (BCI) developed by Michael Porter, and the Global Competitiveness
Index (Global-CI). These indices weight a range of different, essentially macroeconomic, indicators relating to country openness, government, finance,
macroeconomic stability, infrastructure, management, labour, technology, and
civil institutions to produce a global competitiveness score and ranking for 134
national economies.[2] These indices are widely reported in the media world-
wide and have become a fairly established tool for government, institutions
and development agencies seeking to understand national economic performance in relative terms. They have, however, attracted a degree of criticism
(see, for example, Lall, 2001; Huggins, 2003; and Ochel and Röhn, 2006).

The spread of the competitiveness discourse from nations to regions and
also to cities has inevitably led to similar interest in the measurement of
regional competitiveness. Indeed, there has been a proliferation of such indices in recent years and yet, notably, precious little consensus on the best
approach to measurement has emerged. Various 'one-off' attempts have been
made to measure and model competitiveness for European regions (e.g. IFO,
1990; Pompili, 1994; Pinelli et al., 1998; Gardiner, 2003). As a consequence,
the European Commission has placed the analysis of regional competitiveness
at the heart of its ongoing assessment of regional economic performance
(CEC, 1999a; 2000). Similarly, regional competitiveness indicators have
become an established element in the monitoring of regional economic performance in the UK. Indeed, the former Department for Trade and Industry
(DTI) (now the Department for Business, Enterprise and Regulatory Reform,
or BERR) has published sets of separate regional competitiveness indicators
for the standard statistical regions of the UK since 1997 (see for example,
DTI, 2003; BERR, 2008b).

More recently, efforts have been made to develop composite indicators of
competitiveness, following trends in the evolution of national competitiveness
indicators and in response to the desire of policy makers and analysts for
league tables or rankings of regional competitive performance. These league
tables and rankings are typically produced independently of governments, by
a range of different think-tanks and economic development consultancies.

Thus, for example, in the UK various composite indices have been produced by Robert Huggins, firstly as part of Robert Huggins Associates (a private consultancy firm based in Wales), and more recently within the Centre for International Competitiveness based in the University of Wales Institute (UWIC), Cardiff.

Huggins and his colleagues have produced various indices, including the so-called 'World Knowledge Competitiveness Index', which seeks to benchmark a selection of what are deemed to be the world's leading knowledge-based regions in North America, Europe and Asia-Pacific against one another. They have also produced a European Competitiveness Index which ranks the competitive performance of cities and regions across the EU, with a view to 'creating an informed dialogue that can contribute to a policy environment attuned to enhancing the economic performance of Europe's nations and regions'.[3] Finally, Huggins is also responsible for producing the UK Competitiveness Index (UKCI). This was first produced in 2000, with an analysis dating back to 1997 (Huggins, 2000). The methodology developed as part of this study has recently been replicated by other authors, for example, by those seeking to create a competitiveness index for Polish regions (Bronisz et al., 2008). This illustrates that in the development of metrics of competitiveness, the same story of the spread and replication of so-called 'best practice' techniques is discernible.

Similar developments in composite indices have occurred in other parts of Europe. The Bundesländer im Standortwettbewerb (Benchmarking German States) Index (BISW) has been published on a biennial basis since 2001, with rankings dating back to 1986. The report is published by the Bertelsmann Foundation (a private non-profit making think-tank in Germany). The authors of the report assert that the purpose of their indices and analysis is to reflect the fact that 'in this age of globalization, nations are not the only ones competing against each other for jobs and mobile capital – regions and states are increasingly competing intensely as well, both internationally and nationally, for mobile production factors and the associated opportunities for growth and employment'.[4] Benchmarking German states as business sites in the form of state-to-state comparisons is thus designed to stimulate reform of economic and labour market policies and to promote an atmosphere of learning from best practice elsewhere as a way of gaining a competitive edge. This is in line with the reputation of the Bertelsmann Foundation as a strong supporter of supply-side economics, active labour market policies and stricter public sector spending controls. To this end, the Foundation also publishes a public debt monitor, ranking German states according to their levels of public debt, and seeks to stimulate public debate and design the concrete reforms needed to deliver competitive success at state (region) and national level.

Such indices are not, however, unique to Europe. In the US alone, there are at least eight different groups producing competitiveness rankings of states or cities on a regular basis (Fisher, 2005, Greene et al., 2007). For example, the Beacon Hill Institute at Suffolk University in Boston publishes annual reports on state and metropolitan competitiveness across the United States, with the first State Competitiveness Report (SCR) being published in 2001. Furthermore, a plethora of smaller-scale indices have emerged which rank places on the basis of particular measures of competitiveness, such as Richard Florida's 'creativity index' (Florida, 2002). Indeed, in total at present there are at least 33 different empirical studies examining the competitiveness of states, regions or city-regions in the US, UK and across Europe and parts of Asia which are published on a regular basis. These are listed in Table 4.1, alongside some of their defining characteristics.

Index construction: conceptual chaos and methodological pluralism

Conceptions of competitiveness

Those constructing competitiveness indices are confronted with a number of key challenges, not the least of these being what conception of regional competitiveness to base their measurement on and thus what variables to include. A common feature of all indices is their tendency to assert that a multi-dimensional approach is required to the measurement of competitiveness, since 'competitiveness is not an attribute that can be measured directly; all one can do is gauge its nature and magnitude by the shadow it casts' (Kresl and Singh, 1999: p. 1018). On average, the indices listed in Table 4.1 each include 40 indicators, with some including as many as 87 indicators in their analysis. The pluralistic approach to competitiveness measurement is further revealed by the fact that all indices include both price or cost indicators, as well as non-price or quality indicators such as technological capacity and innovation. Measuring everything that might conceivably be relevant seems to be the predominant approach.

All indices typically employ a variety of somewhat ambiguous conceptualizations of regional competitiveness as revealed by their tendency to emphasize microeconomic features of firm performance and productivity, as well as macroeconomic outcomes and place attractiveness (Greene et al., 2007). In short, there is clearly no consensus on what such indices should measure and how they should be constructed.

Some examples illustrate the point. The UKCI takes a broadly macroeconomic perspective. The 2008 index benchmarks 12 regions and some 408 UK localities (i.e. local authority districts and areas). The stated aim of the index is 'to assess the relative economic competitiveness of regions and localities in the UK by constructing a single index that reflects, as fully as possible, the measurable criteria constituting place competitiveness' (Huggins and Day, 2006: p. 60).

The index lacks an explicit theoretical base and framework. However, when defining place competitiveness, the authors explicitly follow the macroeconomic

Table 4.1 Existing indices of regional competitiveness

Name	Author(s) and/or organization	Issuing date/ frequency	Issuing entity type	Geographic focus	No. of entities covered in the latest report	No. of indicators in the latest report	Own survey data included?	Weights applied for overall index
Competitive Alternatives	KPMG	2002, 2004, 2006	PFP	World	137	27	N	NA (special cost model)
European Competitiveness Index	Robert Huggins	2004–	PFP/ PUB	Europe	118	16	N	Non-equal
World Competitiveness Scoreboard	Institute for Management Development (IMD)	Regions included 2003–6	PFP	World	55	246	Y	Equal (implicitly non-equal)
World Knowledge Competitiveness Index	Robert Huggins	2002–5, 2008	PFP/ PUB	World	145	19	N	Non-equal
Annual Zaobao-NTU Competitiveness Ranking & Simulations for 31 Chinese economies	Kang/Giap/ Yam 2006	2006	PUB	China	31	101	Y	Equal (implicitly non-equal)
Best States Ranking	Forbes	2006–	PFP	USA	50	30	N	Non-equal
Bundesländer ranking (Ranking German Länder)	INSM-Initiative Neue Soziale Marktwirtschaft (Initiative for a new social market economy)	2003–	PNP	Germany	16	87	N	Non-equal
Bundesländer ranking Österreich (Ranking Austrian Länder)	Chamber of Commerce Tyrol	2004	PNP	Austria	9	4	N	Equal

(continued on next page)

Table 4.1 (continued)

Name	Author(s) and/or organization	Issuing date/ frequency	Issuing entity type	Geographic focus	No. of entities covered in the latest report	No. of indicators in the latest report	Own survey data included?	Weights applied for overall index
Business Times-NTU Ranking Results on Overall Competitiveness of 35 States & UTs in India	Sen et al 2005	2005–?	PUB	India	35	> 100	Y	Equal (implicitly non-equal)
Competitiveness Ranking of 40 US and 7 Canadian metropolitan areas	Kresl	2004	PUB	USA/CAN	47	3	N	Non-equal
Die Bundesländer im Standortwettbewerb ("Benchmarking German States")	Bertelsmann Foundation	2001– biennially	PNP	Germany	16	50	N	Non-equal
Economic Freedom of North America	Fraser Institute, National Center for Policy Analysis	2002, 2004 6	PNP	USA/CAN	60	9	N	Equal (implicitly non-equal)
Index of regional competitiveness for Finland	Huovari/ Kanga-sharju/ Alanen 2001	2001	PUB	Finland	85	15	N	Equal
Innovative Capacity Ranking: Spanish Regions	Zabala-Iturriagagoitia/ Jimenez-Saez/ Castro-Martinez/ Guitierrez-Gracia	2007	PUB	Spain	17	31	N	Non-equal
Knowledge Worker Quotient: The Top Metros in the Knowledge Economy	Expansion Management	2003– annually	PFP	USA	362	?	N	NA (special model)

Table 4.1 (continued)

Name	Author(s) and/or organization	Issuing date/ frequency	Issuing entity type	Geographic focus	No. of entities covered in the latest report	No. of indicators in the latest report	Own survey data included?	Weights applied for overall index
Metro Area Competitiveness Report	Beacon Hill Institute	2001–5	PNP	USA	50	39	N	Equal
Metropolitan New Economy Index	Progressive Policy Institute (PPI)	2004	PNP	USA	50	16	N	Non-equal
Objective Competitiveness – Ranking of EU Regions	Vicente y Oliva/ Marco Calvo 2005	2005	PUB	Europe	128	63	N	Non-equal
Pinoy Cities on the Rise – The Philippine Cities Competitiveness Ranking Project	Asian Institute of Management	1999, 2002, 2003, 2005	PUB	The Philippines	65	68	Y	Equal
Porträt der Wettbewerbsfähigkeit österreichischer Bundesländer (Austrian States)	Bachner 2005	2005	PUB	Austria	9	8	N	Equal
Regional ranking	INSM-Initiative Neue Soziale Marktwirtschaft (Initiative for a new social market economy)	2006	PNP	Germany	435	47	N	Non-equal

(*continued on next page*)

Table 4.1 (continued)

Name	Author(s) and/or organization	Issuing date/ frequency	Issuing entity type	Geographic focus	No. of entities covered in the latest report	No. of indicators in the latest report	Own survey data included?	Weights applied for overall index
San Diego's Sustainable Competitiveness Index	San Diego Regional Economic Development Corporation/ San Diego Association of Governments	2001, 2005	PNP/ PUB	USA	19	21	N	Equal
Standortradar (Location radar)	Management club Austria	2006–	PNP	Austria	9	26	Y	Non-equal
State Competitiveness Report	Beacon Hill Institute	2001–	PNP	USA	50	42	N	Equal (implicitly non-equal)
State Technology and Science Index	Milken Institute	2002, 2004	PNP	USA	50	75	N	Equal
The Knowledge-Based Economy Index	Milken Institute	2000, 2001	PNP	USA	51	12	N	Equal
The State New Economy Index	2007 Kauffman Foundation, 1999 and 2002 Progressive Policy Institute (PPI)	1999, 2002, 2007	PNP	USA	50	26	N	Non-equal
The Vietnam Provincial Competitiveness Index	US AID/VCC	2005–	PNP	Vietnam	64	64	Y	Non-equal
Toplocaties ('Top locations')	Elsevier (Journal) and Bureau Louter	2002–	PFP	The Netherlands	421	25	Y	Equal

Table 4.1 (continued)

Name	Author(s) and/or organization	Issuing date/frequency	Issuing entity type	Geographic focus	No. of entities covered in the latest report	No. of indicators in the latest report	Own survey data included?	Weights applied for overall index
Top 25 Cities for doing business in America	Inc (Journal)	2004–	PFP	USA	393	4	N	Non-equal
US Economic Freedom Index	Pacific Research Institute	1999, 2004	PNP	USA	50	143	N	Non-equal
UK Competitiveness Index	Robert Huggins	2000–	PFP/PUB	UK	12	15	N	Non-equal
Zukunftsindex Deutschland (Future index Germany)	Prognos and Handelsblatt	2004, 2006, 2007	PFP	Germany	439	29	N	Non-equal

Source: Adapted from Berger and Bristow (2008).
Note: Issuing entity type – PUB = public institutions such as universities or governmental organizations; PFP = private for-profit organization; PNP = private not-for-profit organization.

definition of competitiveness as set out by Michael Storper. Thus they see competitiveness as the capability of an economy to attract and maintain firms with stable or rising market shares in an activity, whilst maintaining stable or increasing standards of living for those who participate in it (Huggins and Day, 2005: p. 43). This framework is reflected in the choice of indicators which focus on human capital, entrepreneurship and innovation as well as overall economic performance.

In contrast, the State Competitiveness Index in the US seems to mix the macroeconomic and Porterian notions of competitiveness to produce an overarching attractiveness-based conception of state competitiveness. Thus, in the view of the authors, 'a state is competitive if it has in place the policies and conditions that ensure and sustain a high level of per capita income and its continued growth. To achieve this, a state needs to be able to attract and incubate new businesses and to provide an environment that is conducive to the growth of existing firms' (Beacon Hill, 2001: p. 5). This asserts the importance of the microeconomic environment within which firms develop and prosper. Indeed, the report states that its approach is inspired by Porter's 'diamond' and his attempt to measure competitiveness as expressed in the Global Competitiveness Report. Thus, higher real gross national product per capita is seen as the ultimate benchmark of higher competitiveness.

Other indices are much less clear as to the conception of competitiveness on which their indices are based. The approach adopted by the Bertelsmann Foundation in the production of the BISW is to refrain from defining regional competitiveness. Instead, a rather broad and essentially empirically driven approach to index construction and data collection is adopted, which focuses on two broad indicators of competitiveness – success and activity.

Inputs, outputs and outcomes

The nature of the interrelationships between the different indicators purported to be essential to regional competitiveness is also somewhat vague across different indices. The great majority of the indices listed in Table 4.1 are composite indices and are thus characterized by a tendency to lump their (often very different) bundles of chosen indicators together in a rather chaotic set of non-linear relationships. This tends to obscure any sense of whether or how competitiveness can be shaped by particular determinants contained within the overall mix and is based on a rather broad-brush understanding of how competitiveness outcomes are produced. This thinking asserts that various 'inputs' (such as innovation and human capital) flow into a range of different revealed 'output' measures (such as economic performance or GDP per capita), which in turn flow into various 'outcome' measures (such as unemployment rates and earnings) (Gardiner et al., 2004).

Thus, for example, the UKCI is constructed exactly along these lines, with a three-factor framework based upon inputs, outputs and outcomes, incorporating 15 indicators. The authors state that this framework is intended to reflect the link between macroeconomic performance and innovative business behaviour, although the nature of this link is not elaborated upon. The input factors used are levels of research and development expenditure; economic activity rates; business start-up rates per 1,000 inhabitants; number of businesses per 1,000 inhabitants; General Certificate of Secondary Education (GCSE) results and NVQ (National Vocational Qualification) level qualifications (both UK post-16 educational qualifications); and the proportion of knowledge-based businesses. These variables are conceptualized as contributing to the output productivity of a region, measured by Gross Value Added (GVA) per head at current basic prices; exports per head of population; imports per head of population; the proportion of exporting companies; productivity (output per hour worked); and employment rates. Finally, the impact or outcome of these different factors is measured with reference to the level of average earnings and the proportion of individuals who are unemployed. The index is thus based on a compromise between what Huggins (2003) has identified as the two main concerns: data availability and explanatory indicators.

In constructing the BISW, the Bertelsmann Foundation takes a more broadbrush approach. The success index (SI) captures indicators of competitiveness outcomes in relation to indicators of employment, income and security. The activity index (AI) focuses on inputs deemed to influence overall outcomes, with these weighted according to econometric analysis of their previous relationship with outcome indicators. Data on a wide variety of over 40 input factors is incorporated into this analysis, including measures of firm activity; educational performance; the spending within regions on higher education and active labour market policies; dependence on subsidies from other regions; employment structures and performance; and overall political stability. Thus, the emphasis on regional attractiveness and the importance of reduced supply-side rigidities is clearly apparent. Results for the different indicators are translated into a scale from 1 (low performance) to 10 (high performance) for ranking purposes. In 2007 the decision was taken to simply report the results of the sub-rankings of each of the indicators in the three key areas of the former AI. The authors state that this was done to draw more attention to the detailed profiles for all 16 Bundesländer analysed, their relative performance, current trends, possible explanations for these trends and to help develop subsequent recommendations.

A smaller number of indices prefer to avoid making any assertions at all about how competitiveness outcomes are determined, preferring instead to present data on a series of different, discrete variables. This is the approach taken by the UK government in its biennial publication of regional competitiveness indicators (see BERR, 2008b). Rather than develop a composite index, this report publishes data on 17 indicators which are 'intended to give a balanced picture of all the statistical information relevant to regional competitiveness and the state of the regions' (BERR, 2008b: p. 5). This approach again, however, implicitly asserts that the concept of regional competitiveness equates with economic performance or 'competitiveness outcomes', which are measured by GVA per capita, labour productivity, investment and output by UK and foreign-owned firms, and exports of goods and services. The remaining indicators are grouped under the headings of the labour market; deprivation; business development; and land and infrastructure. However, no attempt is made to explain how, if at all, these indices relate to one another and how they work separately or in combination to shape competitiveness. The result is the conflation and confusion of the inputs, outputs and outcomes that might constitute regional competitiveness.

Similar criticisms have been levelled against the US State Competitiveness Report (see Fisher, 2005). The plurality of indicators included causes various problems in understanding chains of causality. Some indicators may be construed as being the results of slow growth or low income, such as the unemployment rate and the proportion of households that are uninsured. Other variables are simply correlates with high incomes, such as the percentage of households with telephones, and the prevalence of high-income workers. In addition, a range of outcome variables are included in the mix, such as labour force participation rates, budget surpluses and new firm start-ups. In short, the index mixes causal and outcome variables in a somewhat indiscriminate manner. Furthermore, many variables are missing for many of the states in the analysis.

There are, however, strong commonalities between indices in terms of the measures deemed to be important to include in the overall competitiveness 'mix'. Table 4.2 lists the key variables used in a sample of some of the most widely reported composite indices of regional competitiveness in the US, UK, continental Europe and Asia. The table also indicates in approximate terms the frequency with which variables are included in each of the different indices.

This reveals that as many as 12 out of the 28 dimensions are covered by at least 50 per cent of the indices, with employment/unemployment, innovation and quality of workforce covered by all indices bar one. The importance placed on innovation capacity is particularly evident, with the vast majority of indices including some proxy for innovation activity, whether in terms of R & D expenditure, or patent activity. Furthermore, there is also a strong emphasis on human capital inputs and the importance of education and skills. It follows that productivity also features strongly across the board. There are also, however, some significant differences between these indices. Interestingly only one index covers population growth, and the attitudes and values of regional populations themselves with regard to their standard of living are not covered by any ranking. The World Knowledge Competitiveness Index and the Vietnam Provincial Competitiveness Index have their own particular focus. The World Knowledge Competitiveness Index emphasizes the innovation capacities and knowledge base of an economy. The Vietnam Provincial Competitiveness Index, in contrast, focuses on regional economic governance, such as barriers to local enterprises, and relies heavily on the opinions of local businesses.

Table 4.2 Dimensions covered by sample of regional composite indices

Dimensions/Index name	UK Competitiveness Index	European Competitiveness Index	World Knowledge Competitiveness Index	BISW ('German Länder in location competition')	State Competitiveness Report	Bundesländerranking ('Ranking German Länder')	The Vietnam Provincial Competitiveness Index	Zukunftsindex ('Future Index') Germany	Coverage (%)
Economic performance	x	x		x		x		x	63
Employment/unemployment	x	x	x	x	x	x		x	88
Labour cost				(x)		x			25
Productivity	x	x	x			x			50
High skilled employees (not specified further)						x		x	25
Innovational capacity (patents, R&D expenditures etc.)	x	x	x	x	x	x		x	88
Quality of workforce	x	x	x	x	x	x		x	88
Quality of educational institutions	x			x	x	x	x		63
Political and social stability				x			x		25
Public administration (size of employment share)				x	x	x			38

(continued on next page)

Table 4.2 (continued)

Dimensions/ Index name	UK Competitiveness Index	European Competitiveness Index	World Knowledge Competitiveness Index	BISW ('German Länder in location competition')	State Competitiveness Report	Bundesländerranking ('Ranking German Länder')	The Vietnam Provincial Competitiveness Index	Zukunftsindex ('Future Index') Germany	Coverage (%)
Bureaucratic burden				(x)		x	x		38
Tax burden (corporate tax rate on profits)				x	x	x	x		50
Physical infrastructure (rail, roads, ports etc.)		x			x			x	38
Information and communications technology		x	x		x				38
Entrepreneurship	x				x	x		x	50
Firm performance and solvency				x		x			25
Financial capital, e.g. private equity, FDI			x	x	x	(x)	x		63
Exports (macro-level)	x			x	x				38
Regional demand and purchasing power, earnings	x	x	x			x		x	63
Poverty and inequality				x	x	x		x	50

Table 4.2 (continued)

Dimensions/Index name	UK Competitiveness Index	European Competitiveness Index	World Knowledge Competitiveness Index	BISW ('German Länder in location competition')	State Competitiveness Report	Bundesländerranking ('Ranking German Länder')	The Vietnam Provincial Competitiveness Index	Zukunftsindex ('Future Index') Germany	Coverage (%)
Inflation						(x)			13
Health and sanitation					x	x			25
Ecology					x				13
Quality of life, well-being						x			13
Corruption							x		13
Crime					x	x		x	50
Attitudes and values in general									0
Population growth								x	13

Note: Parentheses used if indices apply special definitions not fitting perfectly.

Data sources and weighting procedures

Having decided on which variables to include, index constructors are also confronted with a range of other methodological issues, including from where to source data, for what regions, and whether and how to aggregate indicators. There is a strong preference for relying on 'hard' or non-survey data from regional or national government statistical offices, which is largely deemed to be both more reliable and more comparable. Nevertheless, as Greene et al. (2007) note, it is not entirely without its problems. In the UK, for example, regional GDP (or more recently, GVA) data has been criticized in the past as being potentially inaccurate and misleading, because of small sample sizes. Of the indices listed in Table 4.2, only one (the Vietnam Provincial Index) includes survey data, seeing it as a good way of assessing certain specific characteristics relating to business preferences.

Decisions as to which regions to include in indices also vary considerably. Whilst most rely upon comparing statistical regions or states within defined jurisdictions (such as within a nation), others embark on broader comparisons across nations, with some (such as the European Competitiveness Index and World Knowledge Competitiveness Index) relying upon European NUTS I regional boundaries. These typically do not correspond to political or statistical conceptions of regions but can vary in size from metropolitan centres (as in the case of Greater London) to entire countries (such as Luxembourg).

There is a rationale at one level for comparing a given region – such as Wales – to all other regions of the national economy, since regions within a national system may compete in different ways and through different means for labour, capital and financial and other resources. However, even within a national system there may be significant regional differences in costs and prices, such that each region's per capita GDP or GVA figure should be adjusted by its own cost or price deflator. In the UK, however, as in many other nations, there are still no consistent and robust time series of regional cost or price deflators, which means that national deflators are typically used (Martin, 2005). The critical point is that decisions about which regions to include and compare in these indices and how these should be defined are largely determined by the availability of comparable data or by what is of interest to the index constructor, and not with any reference to the extent to which these are commensurable spaces competing in commensurable markets.

The aggregation of indicators into composite indices presents a further set of problems, notably relating to the weights to be applied to different indicators. Ideally, weights should be derived from a pertinent theoretical framework, with indicators defined and weighted according to their overall importance. Clearly, in the case of competitiveness, such a framework is noticeable by its absence. At the very least, indices should be transparent in respect to the indicators included, how they are weighted and why, including sensitivity tests of the implications of different weights and/or aggregation techniques.

Not surprisingly, there are considerable differences in the approaches used to weighting indicators in the production of composite indices, with most making value judgements as to what weights to apply to particular variables. Several studies run regression analyses to derive appropriate weights or to choose the correct indicators out of a set of other indicators (as in the case of the BISW report). Others are more ad hoc and pragmatic in their approach to weighting. The US State Competitiveness Index, for example, comprises 42 indicators in eight sub-indices or groupings. As the sub-indices consist of a different number of indicators, the variables implicitly receive different weights. However, the rationale for the inclusion of different indicators and their weighting is not made clear. For example, the 2001 report justified the inclusion of a proxy for salaries in high-technology industries on the basis that high-tech sectors have a 'better' impact on competitive outcomes (Beacon Hill, 2001: p. 14).

Similarly, in the construction of the UKCI each of these three elements (i.e. inputs, outputs and outcomes) is given equal weighting in calculating the overall index, since it is hypothesized that 'each will be interrelated and economically bound by the other' (Huggins, 2003: p. 92). For each measure an index is calculated with an average base for the UK of 100, and the distribution range for each is calculated. A normalization procedure is used to transform each of the variables into a standard logarithmic form so that their distributions are as similar as possible and no single variable distorts the final composite score. All the single indicators are then aggregated into one index for each of the three factors. All values are therefore ranked and expressed in relation to the UK average so that all the values can be compared, leading to numbers lower than, equal to, or higher than 100. The final number for the regional competitiveness index is derived by averaging the normalized scores for all three factors. In order to reflect as far as possible the scale of difference in area competitiveness, the composite scores are then 'anti-logged' through exponential transformation.

Although the authors state that they do not apply any weighting for the final index, due to the different number of indicators under the three measures, outcome factors de facto receive the highest weighting, as only two are used to derive the sub-ranking, compared to seven for the input factors and six for the output measures. No further explanation is given for this rather implicit weighting judgement. In earlier reports greater emphasis was placed on productivity, earnings and unemployment (Huggins, 2003: p. 92). Whilst in all reports equal weights are applied at the level of sub-categories, the number of indicators over time has changed from 6 to 15. This means the implicit weights have changed drastically – in the case of productivity, for example, going down from 33.3 per cent to 5.5 per cent.

Certain indices lack transparency in respect of exactly how weights have been derived and on what grounds. Typically in these cases, vague statements are made regarding the role of expert opinion, own findings or literature analyses as the sources for the respective weights. In Table 4.1, such cases are referred to as 'a priori' weighting decisions.

Interestingly, two indices combine an a priori weighting and regression. In the case of the BISW ranking, a regression is applied to derive the relevant indicators and their respective weights. These weights are then applied at the sub-index level. In aggregating the sub-indices to produce an overall index, an a priori weighting based on surveys is applied. The authors of the Vietnam Provincial Index choose instead to derive appropriate indicators on the basis of a literature review and their own survey data and analyses. The weights for the final aggregation are then derived with the help of a regression analysis. Whilst all of these techniques have their various advantages and shortcomings, the implication is that these decisions can impact on indicators and may produce very different results. However, this fact is scarcely, if ever, acknowledged when these indicators and their results are reported on.

The confusion that characterises the construction of competitiveness indicators and rankings is perhaps not surprising. It clearly mirrors the confusion which surrounds the meaning of competitiveness itself and the absence of a coherent theoretical framework capable of providing appropriate guidance on the selection of appropriate variables and their interrelationships. As a consequence, index crafters inevitably have to rely very heavily on expert judgement and ad hoc empirical analysis (Rouvinen, 2001).

The utility of competitiveness indices

Composite indices of competitiveness appear to have some utility. They are certainly widely reported in the media and help to satisfy a need for data on regional economic performance emanating from various regional development agencies, heads of regional government, investment agencies and chambers of commerce (see, for example, Beacon Hill, 2007). The league table format is inevitably seductive for those analysts and practitioners keen to absorb 'quick and dirty' comparative measures of regional economic performance, or to tell a positive 'success' story about their particular region which may be 'top of the league' or perhaps 'catching up on' purported rivals. In the UK, the results of the Competitiveness Index have, for example, been used to highlight the rather glacial progress being made in addressing the continuing north–south divide in economic fortunes across UK regions and to recommend a more active and targeted regional development policy (e.g. Thornton, 2006). The publication of the first ranking in 2000 captured particular media attention because it used comparative data from the *World Competitiveness Yearbook* to suggest that the level of competitiveness in Wales, one of the poorest UK regions, was similar to that of nations such as Chile, Hungary and Israel (Huggins, 2000). The headline-grabbing potential of these tables and rankings is thus clear, and to the extent that they produce stories of performance which accord with other indicators, they clearly may have some value in reinforcing understanding of differential spatial patterns of economic performance.

For example, since its inception the UKCI has produced fairly consistent results in terms of the rankings of UK regions (as shown in Table 4.3, which compares the results for 2008 with those of 2006). This reveals a high degree of inertia in the regional rankings, with very little change evidenced over this period. Indeed, the top five positions remain unchanged, with the list headed by the 'Big Three' regions of London, the South East of England and Eastern England – often referred to as the 'winners' circle'. The UK's most 'uncompetitive' regional economies remain the North East of England, Wales, Northern Ireland and Yorkshire and Humberside, which clearly run the risk, according to this data, of being stigmatized as 'failing' because of their own deficiencies, when in fact, of course, their problems may lie, in part, in broader structures (see Chapter 6).

This inertia in rankings suggests that the challenge facing policy makers seeking to move their regions up the table is a considerable one. In other words, relative regional economic performance is not quickly improved, especially when every other region is trying to develop and grow at the same time. The scores behind the rankings do, however, change and over the period between 1997 and 2008 the gap between the worst and best performing regions has fallen from 40 'points' in 1997 to 29.4 'points' in 2008. This implies a narrowing of regional disparities in the UK, over this time period at least.

One way to assess the robustness of this index is to examine the relationship between the ranking results it provides and average regional GDP growth rates from 1997 to 2006. This reveals a strong correlation such that the most competitive regions according to the UKCI are those which have the

Table 4.3 UK Regional Competitiveness Index 2008 (UK = 100)

Rank 2008	Region	UKCI 2008	UKCI 2006	Rank 2006	Change in rank
1	London	112.5	113.9	1	0
2	South East	109.7	110.5	2	0
3	Eastern	105.6	106.0	3	0
4	East Midlands	97.7	96.1	4	0
5	South West	95.0	94.9	5	0
6	North West	94.5	92.3	8	+2
7	West Midlands	94.4	92.7	7	0
8	Scotland	94.3	94.2	6	−2
9	Yorkshire and the Humber	89.6	90.5	9	0
10	Northern Ireland	88.8	88.0	10	0
11	Wales	86.8	86.7	11	0
12	North East	83.1	84.2	12	0
UK	UK	100	100		

Source: Huggins and Izushi (2008)

highest rate of GDP growth (Berger and Bristow, 2008). This is perhaps not surprising, given that GDP forms an important part of the index itself. The correlation is much weaker and not significant when the ranking is compared with unemployment rates over the same period. The index thus functions as a rough proxy for regional economic performance, although in this regard its added value over and above the use of GDP or GVA data is not entirely clear. In fact, it arguably has *less* value, inasmuch as the index number format creates a curious abstraction in the values under consideration. Composite rankings fail to illuminate the scale of deficiencies in key indicators in each region relative to comparators. Indeed, a stable rank ordering of regions could conceivably be consistent with progressive widening or narrowing of regional disparities in per capita GDP, and the more aggregated or composite the index, the more ambiguous the comparisons are likely to be.

Herein lies a major problem with competitiveness indices: they are a very poor guide for public policy. Their tendency to conflate and confuse different input, output and outcome indicators makes it very difficult to assert what particular remedial policy interventions within regions are necessary to achieve the desired improvement in outcomes. One might reasonably ask, what does the UKCI *mean* when it states that the South East of England is currently 16.6 'points' more competitive than the North East of England? How can policy makers or development agencies understand and act on this, especially when the composite bundling of inputs, outputs and outcomes makes it almost impossible to diagnose precisely from where supposed 'failings' in competitive performance stem and what levers can be pulled to remedy them?

The weaknesses of composite indices may imply a preference for a simple reliance on single, revealed measures of competitiveness, such as productivity or output per head, but this is not ideal either. The simple equation of productivity with competitiveness is problematic, since improved productivity is not necessarily a basis for employment growth and thus broader improvement in regional prosperity and living standards, another measure of regional competitive performance. Indeed, London has enjoyed the highest rates of productivity growth across the UK in recent years, but its employment growth has been below the national average. Similarly, the high salaries earned in certain key sectors, such as financial services in London, inflate its productivity level and give a somewhat artificial measure of its 'success' over and above that of other UK regions (Massey, 2007a). Furthermore, whilst temporal trends in the growth of relative productivity or GDP may provide measures of revealed competitiveness, they do not of themselves help to identify the sources of that competitiveness (Martin, 2005).

This points to the wider deficiencies in standard 'revealed' measures of competitiveness or regional economic performance. Conventional league tables of competitiveness, GDP and productivity typically conceal more than they reveal about spatial inequalities in well-being and their causes. Morgan (2004a) observes that regions can appear at similar points in narrowly economistic league tables, but the same regions can be very different in terms of real quality of life. For example, the regions of the Italian Mezzogiorno are as poor as Wales in terms of income, but they do not suffer from the same debilitating rate of limiting long-term illness, partly because they have access to a much healthier diet. Poor health is both a cause and a consequence of a weak labour market in Wales because high rates of limiting long-term illness are part of the explanation for the debilitating levels of economic inactivity in the region. Indeed, in all regions the objectives of economic prosperity, a fair society, environmental sustainability and quality of life cannot be regarded as mutually exclusive. However, neither are they entirely consistent, and managing the trade-offs and tensions between them requires considerable political statecraft and challenging deliberations about the relative merits of individual choice versus the collective good, and the importance of values as well as material gains.

Competitiveness metrics also ignore other activities that influence the well-being or liveability of regions, including non-monetary activities such as the informal unpaid economy of domestic labour, of caring and sharing, nurturing of the young, volunteering and mutual aid. Yet according to Henderson (1999), this activity represents some 50 per cent of productive work and exchange in all societies. Furthermore, conventional measures of GDP or GVA make no distinction between transactions that add to well-being and those that diminish it. Thus GDP includes as positive additions to the index the production of economic 'bads' such as exploitative destruction of the natural environment and pollution, as well as the money spent fighting the breakdown of social structures, maintaining prisons and the criminal justice system and so on. Thus, for example, increased fuel consumption and concomitant CO_2 emissions are equated with an increase in 'standard of living' – a perversity which can only seemingly encourage and incentivize further production and more waste (see Jackson, 2004). Moreover, these metrics say nothing about distributional concerns which, as Massey (2007a: p. 61) observes, ought not to be simply 'a caveat to an achieved success, but part of the criteria of success itself'.

The point is that these differences clearly matter and yet are masked by the crudeness of conventional measures of competitiveness and economic performance. The result is badly informed policy. Indeed, emergent geographies of well-being appear to challenge the accepted simple geographies of 'success' portrayed by competitiveness league tables (Massey, 2007a). The most 'successful' and competitive economies in the UK are often those characterized by the lowest levels of quality of life or social well-being, inasmuch as they experience higher levels of crime, lower levels of health, greater problems of housing and service affordability and lower environmental quality. In contrast, less 'competitive' regions and cities are often characterized by a more complex and place-specific mix of social and environmental positives and negatives. The 'winners' circle' in and around the South East of England looks much less attractive when account is taken of its high levels of 'ill-being' in terms of crime, deprivation, high house prices and a seriously overburdened transport system. The dynamics of competitive success in the South East appear to be creating a highly centralized economy, large swathes of which are unaffordable, as well as generating ever-increasing problems of commuting and congestion, thereby undermining environmental sustainability goals (Local Futures Group, 2006).

Metrics that matter

The limitations of competitiveness indices and narrow metrics of economic performance are beginning to be more openly debated. In particular, there is growing interest in the development of a calculus of happiness as a broader assessment of utility than one based simply on exchange value (e.g. Layard, 2005). This has some merit in highlighting the fissure between economic objectives and life satisfaction in developed countries, but suffers some limitations inasmuch as people may report themselves as being happy with their lot when they may in fact be undernourished, poorly housed and experiencing little or no prospect for improvement (Jackson, 2009). In this regard, there appears to be greater value in the work of Amartya Sen (1999) and of Martha Nussbaum (2000), who argue for a concept of living standards based on the capabilities of people to flourish in any given place or context. Nussbaum builds on Sen's 'development as freedom' work by identifying more specifically what it means for humans to flourish and thus what should be the list of 'central human capabilities', whilst acknowledging that these must be understood as existing within the clearly defined limits set by the finite nature of ecological resources and the scale of the global population.

Clearly, as yet we may be far away from the emergence of new global ideals around the collective social improvement of the opportunities or 'capabilities' of all individuals, the pursuit of happiness, sustainability and overall well-being – the moral norms that unite us and provide constraints on the utility-enhancing choices nations may make. However, the search for a metric which treats progress and well-being as one of its premier indicators of development is gathering momentum at the international scale. For example, the Stiglitz–Sen Commission on the Measurement of Economic Performance and Social Progress is currently reviewing a range of alternative metrics of development with a view to establishing how to improve the calculation of GDP; how to incorporate new measures of economic, social and environmental sustainability into national accounting; and how to devise fresh indicators for assessing national quality of life.[5] A similar debate is gaining momentum at national level with a developing clamour for data on 'real wealth' in respect of personal and social well-being and community resilience – or 'the things that really matter' (New Economics Foundation, 2009). Regions and cities are also increasingly in the vanguard of the development of new, more holistic indicators of sustainable welfare, liveability and progress (see, for example, Chapter 5). The ability of these metrics (and the broader conceptions of prosperity and development on which they depend) to challenge conventional wisdom clearly remains to be seen.

Conclusions: 'jiggery-pokery' and the utility of competitiveness metrics

The growth in interest in competitiveness indices across the world's nations, regions and cities cannot be underestimated and it is fair to say that in many countries they have become an accepted part of the regional data-reporting landscape, a prosaic, commonplace occurrence helping to tell the story of which regions are 'winning' the competitiveness game and which are lagging behind. But their use and acceptance, as with the concept on which they are based, has run far ahead of any detailed critical interrogation of their conceptual basis, methodological construction and utility as a tool for economic and policy analysis.

This chapter has sought to open up this debate by beginning to problematize regional competitiveness indicators, their function, theoretical basis and robustness. This has revealed a number of significant problems. Fundamentally, and perhaps not surprisingly, the indices demonstrate very clearly the innate lack of consensus as to what regional competitiveness means and how to operationalize and measure it in practice. The sample indices explored here typically borrow from and use all the different definitions of competitiveness explored in Chapter 1, and tend to conflate inputs (such as the knowledge intensity of businesses), outputs (such as productivity), and outcomes (such as income growth and measures of prosperity). Moreover, there is no clear sense of what should be the chain of causation (if indeed there is one) between these different variables. This confusion is further evidenced by the lack of consistency in which variables to include and how they should be weighted and aggregated in the overall analysis. The absence of a coherent competitiveness framework capable of offering any guidance to index crafters means that they are largely required to depend upon expert or subjective judgement, data availability and rather ad hoc empirical analysis. In short, the construction of competitiveness indices falls far short of being an exact or agreed science.

Within the breadth of approaches, a broad dichotomy is discernible between analyses which consist of the reporting of a series of separate indices, and those which seek to develop composite indices, where a range of input, output and outcome variables are measured and aggregated to form a single overarching measure of competitive performance. The increased popularity of composite indices reflects the growing urge to benchmark or rank the comparative performance of one region against another. Composite indices and league tables simplify complex measurements constructs and reveal a competitiveness story, and thus have considerable political and media appeal. However, in so doing they implicitly assume that regions are clearly defined, coherent, atomistic and bounded entities with quantifiable (exclusively economistic) attributes that are in their possession and under their control. This does not seem to square with the growing acceptance of regions as rather more open, diverse and heterogeneous entities with fluid relationship boundaries. Furthermore, such indices and league tables necessarily pull us into the trap of making direct comparisons between regions which are often very different and not necessarily playing the same competitiveness game on the same playing field and with the same conditions, rules and numbers of players and so on. These subtle nuances are lost in the aggregation process and unlikely to be reported in the headlines, however, thus creating potentially misleading and undoubtedly negative impression of those regions deemed to be at the bottom. Decisions about which regions to choose for benchmarking and indicator construction are often made on the basis of data availability, or pragmatically on the basis of regions within a defined national-space economy, when it may be more insightful to compare against regions with similar histories and development trajectories, perhaps in different national contexts. Without this, competitiveness league tables might as well be comparing apples with pears.

The value of such indices is therefore questionable beyond their purpose in reminding us of the continued success of particular regions and paucity of others, and thus in encouraging policy makers to indulge in the hubris of place promotion (Greene et al., 2007). Rather, their value can be construed as being of more symbolic than tangible importance. They help to elucidate a kind of revealed competitiveness or a perceived level of performance on an agreed set of proxy indicators, and this is what ultimately appears to matter more to policy makers and business investors than their ability to accurately measure actual economic performance and provide clear guidance as to how the elusive notion of competitiveness can be achieved. Such indices can serve a useful purpose in highlighting differences between regions in particular economic circumstances. Thus, the business community uses ranking as a tool to determine investment plans and to assess locations for new operations (Ochel and Röhn, 2006), whilst governments and policy officials use them to identify particular areas of an economy's weakness or make the case for particular public policies or strategies for inducing growth (Fisher, 2005; Dunning et al., 1998). It is of course questionable whether league tables and rankings are specifically necessary for such a purpose.

The appeal of these indices and rankings seems to outweigh any sense of concern over their accuracy or reliability for providing specific policy guidance. The evolutionary 'survival of the fittest' basis of the regional competitiveness discourse strongly appeals to the stratum of policy makers and analysts who can use it to justify what they are doing and/or to find out how well they are doing relative to their 'rivals'. The attractiveness of the competitiveness discourse is thus partly a product of the power of the mathematical nature of the economics discipline and the business strategy literature from which it emanates. This creates an innate impartiality and technicality for the market outcomes (such as competitiveness) it describes (Schoenberger, 1998).

In part, the growth in competitiveness indices and benchmarking is a product of the growing audit culture which surrounds the neoliberal approach to economic governance in market economies. In an era of performance indicators and rankings, it is inevitable that regions and cities should be compared against each other in terms of their economic performance. Public policy in developed countries experiencing the marketization of the state is increasingly driven by managerialism which emphasizes the improved performance and efficiency of the state. This managerialism is founded upon economistic and rationalistic assumptions which include an emphasis on measuring performance in the context of a planning system driven by objectives and targets (Sanderson, 2001). This is closely intertwined with assumptions about the increasingly global nature of economic activity. Thus, as the view that economies are self-contained and self-regulating systems has been replaced with the view that economies are locked in unyielding international competition, a new relationship between the economy, the state and society has emerged in which their distinctive identities as separate spheres of national life are increasingly blurred. As a result, there is increasing pressure 'to make relationships based on bureaucratic norms ... meet the standards of efficiency that are believed to characterize the impersonal forces of supply and demand' (Beeson and Firth, 1998: p. 220; cited in Greene et al., 2007). This in turn leads to an increasing requirement for people, places and organizations to be accountable for their performance and for success to be measured and assessed. In this emerging evaluative state, performance tends to be scrutinized through a variety of means, with particular emphasis placed upon output indicators. This provides not only a means of lending legitimacy to the institutional environment, but also some sense of exactitude and certainty, particularly for central governments, which are thus able to retain some 'top-down', mechanical sense that things are somehow under their control (Boyle, 2001).

In this regard, it does not really matter that competitiveness indices are flawed and a poor basis for specific policy action. Their significance and utility lies in the broader effort of co-opting regions into support for competitiveness agendas, creating a sense of shared purpose and cognitive convergence on key policy ideas and tools, if not necessarily of precise decisions and actions. Competitiveness indices and benchmarking practices help to reinforce the notion of a competitive game, of having to act to maintain external attractiveness for investment, as well as undertaking supply-side reforms to effect improvements in business productivity and performance. They are ostensibly a technology of government and a mechanism by which key institutions in particular act to promote and disseminate the competitiveness rationality. Cammack (2006: p. 10) describes the 'rapid spread of surveillance, benchmarking and peer review through coercive or cooperative supranational mechanisms and close co-ordination between national competitive authorities' and explains how indices have progressively broadened in scope over time. In so doing, he concludes, 'behind all the jiggery-pokery that this entails, the principal purpose of the annual league tables is to support national reformers, aiding and abetting the social/socio-psychological process of "locking-in"' (p. 12).

However, by concentrating on the mere statistics of GDP and other conventional monetary indicators, competitiveness indices fail to enable us to distinguish between the qualitative aspects of growth: healthy or unhealthy growth, temporary or sustainable growth. We do not question what growth is actually needed or what is required to actually improve the quality of our life. To put it another way, competitiveness indicators provide an inordinately narrow, economic metric of development which focuses our attention on inputs or *means* to development (such as income growth and productivity improvement) – things which are *instrumentally* significant – rather than on longer-term policy *ends* or outcomes (such as health, well-being and a sustainable natural environment – things which are *intrinsically* significant (Morgan, 2004a). As such, the search is on for new, more progressive metrics of development and progress, and it is a search which, although challenging, is clearly gathering pace and momentum.

5 Resisting or restating competitiveness?

Variation, recontextualization and the role of the regional state

Introduction

As Chapter 3 has demonstrated, it is increasingly acknowledged that regional development policy exhibits a strong tendency towards convergence and institutional isomorphism, such that regions across the world appear to be pursuing 'identikit' development strategies. This serial reproduction of ideas and strategies has been explained by the power of the discourse which unifies them – the discourse of competitiveness. The hegemonic status of competitiveness, coupled with its particular prescriptions for understanding and securing regional economic success and the transmission of its ideas by particular players and through key networks, plays a powerful role in propagating the sameness of ideas and approaches that is evident.

However, Chapter 3 has also suggested that whilst there are strong pressures for convergence around the discourse of competitiveness and the key policy levers and tools deemed essential to its delivery, this convergence in 'talk' may not necessarily result in convergence in decisions and actions. Indeed, the emergent cultural political economy (CPE) approach outlined in Chapter 2 points to the dangers of assuming that dominant strategic approaches are automatically replicated and reproduced within the contingencies of a particular economy. Instead, significant scope exists for dominant economic imaginaries such as competitiveness to be subject to processes of variation. This may occur for a number of reasons, including their incomplete mastery, their skilful adaptation or 'recontextualization' to meet specific circumstances, or as a result of new challenges or crises which often produce profound strategic disorientation and propagate a proliferation of alternative discourses (Jessop and Oosterlynck, 2008).

The purpose of this chapter is thus to identify and explore the range of different variables that may shape the transmission of policy ideas around regional competitiveness into tangible interventions and models of economic governance in practice. In particular, it focuses on exploring the relative importance and agency of the key institutions of state power at the regional level and their capacity to express, represent and mobilize a coherent, collective vision amid competing interests and development agendas, as well as their power to decide on and deliver transformative action. This directly responds to the call by Jones (2008) for a more sophisticated understanding of the relationships between the processes of state rescaling and the spread of key economic imaginaries such as competitiveness. To illustrate the changing competitiveness agenda being pursued by the devolved regional government reference is thus made to the case of Wales in the UK. In so doing, the chapter seeks to address key questions around whether and how regional governments have the potential to modify and vary the dominant competitiveness discourse, and the extent to which their scope for action is constrained. This, in turn, allows for consideration of whether recontextualization of the competitiveness discourse is likely to result in constrained heterogeneity in strategic approaches and outcomes and thus its continued dominance (or 'requisite variation'), or whether more powerful scope exists for variation in the form of resistance.

The chapter begins by utilizing the CPE framework and broader political economy literature to identify the range of variables which are likely to impact on and effect variation in competitiveness strategies and policy approaches between regions. It then proceeds to establish more specifically the selection and mediation role played by devolved regional governments, before exploring the impact that the establishment of regional government has had in the development of competitiveness strategies in the case of Wales. The chapter concludes by examining the implications for the continued dominance of the regional competitiveness discourse.

Cultural political economy and the 'recontextualization' of competitiveness

In outlining the tendency for regions to deploy the same competitiveness strategies and policy tools, Chapter 3 introduced the paradox of uniqueness which highlights the contradictory forces within which regional economic development strategies increasingly have to be formed. Whilst processes of benchmarking and locational competition provide a powerful stimulus for convergence as regions increasingly compare their policies and institutions to so-called 'best practice' exemplars, the competitiveness discourse, as Fougner (2006: p. 182) observes, 'stimulates divergence of a kind, as it becomes of paramount importance for [regions] to stand out as different *and* better in order to appear attractive in the eyes of capital'. Moreover, and following Radaelli (2008: p. 250), it does not necessarily follow that regions wedded to the same competitiveness discourse will necessarily pursue strategies and policies in the same way and with the same effect, not least because of different socio-political contexts or 'institutional legacies, state traditions, and the dominant legal culture'. Furthermore, there may be implementation gaps, or challenges in securing effective replication in strategies on the ground (as Markey et al., 2008).

In addition to these bottom-up or endogenous drivers for policy variation, there may also be good reason for expecting a range of top-down influences to effect divergence in regional competitiveness strategies as they are played out on the ground. Not least among these relates to the contingent variation that resides within the broader milieu of neoliberalism and globalization. Cox (2004) argues that, as well as there being some resistance within different contexts to the turn to neoliberalism, the challenge and power of globalization may have been somewhat overestimated, since capital has both homogenizing and differentiating tendencies. He goes on to posit that the challenge is to critically scrutinize and understand 'the tension between the universalizing on the one hand and the particularizing on the other' (p. 191). Likewise, Cerny (2007) draws attention to the complexity and plurality of the globalization discourse and states that national and regional differences belie its homogeneous appearance. Similarly, Peck and Tickell (2002) assert that the transformative and adaptive capacity of neoliberalism has been repeatedly underestimated, and argue for a closer reading of its historical and geographical (re)constitution and of the variable ways in which different 'local neoliberalisms' are embedded within wider networks and structures of neoliberalism.

This begs numerous questions around the variation processes that may occur as the competitiveness discourse is transmitted across regions. These include, firstly, the extent to which regional competitiveness is differentiated according to policy choice, error, adaptation and/or recontextualization processes; secondly, what the main determinants of these processes are; and thirdly, how significant a challenge (if any) to the hegemony of competitiveness is the variation that they produce. Indeed, key proponents of the CPE framework have made a plea for closer interrogation of the discursive and extra-discursive mechanisms that result in processes of variation and innovation, as well as deeper scrutiny of the relationships between local and extra-local sites and scales, such as national and supranational governments and policy programmes and their framing of local possibilities (Jessop and Oosterlynck, 2008). In so doing, these authors have also drawn attention to the historical specificity and materiality of economics and the potential significance of path-dependent legacies in different spatial contexts. This finds resonance with Lawton-Smith et al. (2003: p. 870), who, in an analysis of European policies designed to improve the competitiveness of regions, conclude that 'while European policies, which prioritize the region as the scale of delivery, empower sub-national responses, particular political-economic histories shape how those responses are translated into policy'.

In an analysis of the regional innovation policies introduced by different Spanish regional governments from the mid 1980s onwards, Sanz-Menendez and Cruz-Castro (2005) observe that the variables underlying both policy choices and the extent to which they vary from one region to another may be usefully summarized as ideas, interests and institutions. This clearly accords with the work of Hirschman and the CPE approach contained in Chapter 2, which has highlighted the significance of both the elasticity of the idea of competitiveness as well as the power of the interests promoting it.

The CPE approach draws particular attention to the ways in which material processes, such as financial difficulties and economic crises associated with developments such as de-industrialization, are interpreted discursively (Jessop, 2004). The approach suggests that a discursive interpretation of the character of the particular crisis takes place and the choice of certain strategies over others is formulated against this interpretation. Furthermore, it also implies that whilst certain 'meta-discourses' such as 'globalization' or the 'third way' can reinforce and help to legitimize key economic imaginaries such as the discourse of competitiveness, this discourse may in turn be recontextualized through various 'micro-discourses' in specific settings (Dannestam, 2008).

Observing that the interventions of sub-national government are much more diverse than the prevailing view about the convergence of regional policies may imply, Sanz-Menendez and Cruz-Castro (2005) highlight the relevance of mobilized interests when they are concentrated in the region. Indeed, they demonstrate that changes in policy orientation are particularly dominant when those interests play a role in the administration of such policies. The importance of understanding the capacities and strategies of social forces mobilized behind competing imaginaries and interests within regions is also highlighted by Krueger and Savage (2007), who state that, to function effectively, regions and city-region states are increasingly required to marshal the resources to support a wide variety of needs and interests relating to sustainability and 'liveability' (or quality of life), and not just the competitiveness-orientated concerns of a narrow group of 'elites'. As such, they are likely to be characterized by multiple class and political alliance formations and a range of conflicts around the management of collective consumption between different sets of interests or 'winners' and 'losers' (Ward and Jonas, 2004). This raises further questions, however, as to how key discursive constructs emerge and are contextualized in relation to binaries or alternative concepts and ideas from different interests, and to what extent these represent a force for resistance and change through transformation into concrete regional politics, or are simply a challenge to be accommodated by competitiveness agendas and interests.

The CPE approach also highlights the importance of how key economic imaginaries are selected and retained in different settings through the material practices of institutions, organizations and actors. It regards state power as subject to a changing balance of forces, the result of which is a constant struggle relating to its identities, subjectivities and interests. As such, a range of economic, political and intellectual forces work together to encourage the development of new structural and organizational forms that help to institutionalize the boundaries and temporalities of different economic imaginaries in an appropriate spatial-temporal fix. This, in turn, can displace and/or defer capital's inherent contradictions and crisis tendencies. This is because, the greater the range of sites (both horizontal and vertical) in which resonant discourses are selected, the greater is the potential for them to be retained through their effective institutionalization and integration into processes of structured coherence and durable compromise. Furthermore, such processes of strategic selectivity of specific organizational and institutional orders are most powerful where they operate across many different sites and can work to promote complementary discourses across wider policy and social ensembles. Thus, in order to understand which variations and innovations become selected and take hold in regions, it is necessary to understand the factors shaping their discursive resonance with particular actors and social forces and/or their reinforcement through a variety of structural mechanisms (Jessop and Oosterlynck, 2008).

More specifically in relation to competitiveness, this means exploring what it is in the logic and mindset of existing institutions and actors that resonates with the competitiveness discourse; which actors promote different discourses; and which technologies they employ. In particular, it implies the need to examine the role of state rescaling and the reshaping of state spatiality, which, as Jones (2008) and Lagendijk (2007) have suggested, are inherently related to, yet may act variously to reinforce or even withstand, the selection and spread of dominant economic imaginaries. This focuses attention on emergent sites of power at the regional scale which have risen in importance alongside the spread of regional competitiveness. This includes both the formal institutions of regional government and the increasingly wide range of actors engaged in decision-making processes in the region who may form key coalitions for the exercise of 'discursive power', if not necessarily governing capacity (Dannestam, 2008). Important questions thus become whether there is resistance within the organization of regional government and in various actors in the region to the spread of competitiveness; where, if any, are the discursive lines of battle; and what impact this has on adopted strategies. It is to the specific role of regional government that this chapter now turns.

Regional 'state spaces' and the selection and mediation of competitiveness

Across the developed world, increasing faith is being placed in the spatial and territorial reorganization of the national state apparatus to deliver public policy and to reinvigorate economy, state and broader civil society. The devolution of powers downwards from the national state to sub-national, particularly regional, forms of government and governance, has indeed become a global trend, with increasing emphasis being placed upon the potential for decentralization to achieve an economic dividend or gain (Rodriguez-Pose and Gill, 2005). Regions have become actors linked together in various networks which increasingly bypass the nation-state, which has become 'hollowed out' as a result (Jessop, 2001). In addition to these scalar shifts in state power, regulatory capacities have also been shifted 'outwards' to a range of non-state organizations and institutions which have been incorporated into processes of governance. The growing emphasis on governance, and networked forms of regional governance in particular, reflects the growing importance of institutions of civil society and of cultural forms in supporting the functioning of the economy. This is resulting in the emergence of new forms of economic governance designed to mobilize available institutional and productive sources and to develop a coherent sense of overall economic identity (Jones et al., 2005a). This is also leading to the emergence of shifting 'state spaces' (Brenner et al., 2003), and has created a highly pertinent dialectic whereby new sites for the institutionalization of neoliberal forms of economic governance are being formed which, at one and the same time, are upheld as intrinsically providing space within which existing trajectories of economic governance can be altered and policy variation effected. As Walker (2002: p. 5) succinctly puts it, 'the beauty of devolved governance is that it can do things differently'.

The precise role played by these new regional state spaces in mediating both the selection of competitiveness and the processes of contingent variation and/or resistance has, however, been relatively under-explored to date. Yet it may clearly be very significant, not least because of the dual role regional state spaces have the capacity to play in propagating the mantra of competitiveness. On the one hand, as instruments of economic governance they clearly have the capacity to promote the development of competitive firms within their jurisdictions, and to devise and implement a range of different strategies and policy tools to this effect (see Chapter 3). On the other hand, and in accordance with the conception of competitiveness as akin to place attractiveness (see Chapter 1), they are at one and the same time themselves competitive entities (to a greater or lesser extent), subject to intensifying pressures to act in an entrepreneurial fashion, gain first-mover advantage in competitions for key investments, events and resources, and to 'sell' the attractions of the region they are deemed to govern and represent. Inasmuch as regional state spaces are instruments of neoliberal economic governance, they may thus be deeply inscribed with a competitiveness imperative. They are both competitive spaces *and* spaces of competitiveness.

In this regard there are clear parallels with the debates around the 'entrepreneurial city', which is the label increasingly used to describe the tendency for cities to focus on the pursuit of 'innovative strategies intended to maintain or enhance [their] economic competitiveness vis-à-vis other cities and economic spaces' (Jessop and Sum, 2000: p. 2289). Regions, like cities, in this discourse have been constructed as relevant political categories or scales for the regulation of political-economic processes. As will become clear, however, this does not necessarily mean that the 'region' as defined for competitiveness purposes is necessarily equivalent to the region aligned with the scope and objectives of regional government.

Certainly, a wealth of international evidence and opinion pays testament to the increasing dovetailing of the discourses of regional devolution and decentralization with those of competitiveness and economic success (see also Chapter 2 for discussion of the complex interrelationships between discursive constructions of the region and of competitiveness). Rodriguez-Pose and Gill (2005) demonstrate how proponents of regional devolution in a variety of nations, including Italy, the UK, Spain, the US and Mexico, have increasingly deployed economic discourses around competitiveness, innovation and regional capacities to adapt to globalization and economic change in their arguments for greater decentralization of power. Indeed, the devolution of powers and responsibilities to regional institutions, whether democratic or more narrowly administrative, appears to be given added strength by the arguments contained within the regional competitiveness discourse. There is clear political capital to be gained from highlighting endogenous capacities to shape economic processes, not least because it helps to generate the sense of regional identity that motivates economic actors and institutions towards a common regional purpose (Rosamund, 2002). Furthermore, the regional competitiveness discourse points to a clear set of agendas for policy action over which regional institutions are likely to have some potential for leverage – agendas such as the development of university–business relationships and strong innovation networks.

In the UK in particular, the programme of regional devolution set in motion from 1997 onwards was couched ostensibly in terms of enabling regions to unlock their localized competitive strengths: 'the Government believes that a successful regional and sub-regional economic policy must be based on building the indigenous strengths in each locality, region and county. The best mechanisms for achieving this are likely to be based in the regions themselves' (HM Treasury, 2001: p. vi). Similarly, in Germany over the course of the 1990s, the once-dominant Fordist-Keynesian project of promoting spatial solidarity was gradually reformulated into an emphasis on regional economic growth and competitiveness alongside administrative self-reliance (Brenner, 2000). In this way, the discourse of regional competitiveness helps to provide a way of constituting regions as legitimate agents of economic governance. In turn, regional devolution provides an essential means by which the national state changes shape, delineates and strengthens regional spaces of global accumulation and thus institutionalizes and spreads the competitiveness agenda (Brenner, 2000).

However, there are strong grounds for questioning whether this interdependence between devolution and regional competitiveness is as mutually supportive, stable and fixed as it might at first seem. In other words, the processes associated with regional governance may not necessarily present an institutional structure that in and of itself materially supports the selection and retention of the discourse on competitiveness. Fundamentally, this reflects the fact that the devolution process is a complex and variegated product of a range of different interest conflicts between the actors involved at national, regional and increasingly sub-regional level (Rodriguez-Pose and Gill, 2005; Harrison, 2008b). Accordingly, and following the understanding that regions are both spaces of competitiveness and competitive spaces, regionalization and devolution processes are similarly the outcome of both economic developments and political mobilization and rescaling processes.

First and foremost, these processes are often strongly steered by national governments, such that the state, undeniably, plays a crucial role in enabling or restricting the devolutionary process, defining how it operates, and even governing its outcomes (Harrison, 2008b). This 'centrally orchestrated region-alism' (Harrison, 2008b) is revealed in a number of ways, including the tendency to create structures (such as Government Offices for the English regions) which are mandated with limited powers, capacities and resources and so which, in effect, become mechanisms to facilitate the regional intervention of central policies rather than to promote autonomous regional action. Similarly, other bodies such as Regional Development Agencies (RDAs) (as in the case of the English regions) may be created which are state sponsored and typically have 'responsibility without resource' (Morgan, 2002; Pearce and Ayres, 2009), whilst regional governments may be mandated with constrained remits and limited financial, fiscal and regulatory autonomy such that they become perceived as another layer of regional bureaucracy with only the power to decide rather than the power to effectively innovate or transform (Morgan, 2004a; Markey et al., 2008). For example, in the study of Spanish regions and their innovation policies referred to above, Sanz-Menendes and Cruz-Castro (2005) find clear evidence to suggest that regional preferences towards a policy change or reorientation are much less likely to succeed, the weaker the powers of regional government.

As such, regional devolution may be regarded as less of an exercise in rolling back the frontiers of the national state, and rather a reflection of the state's re-territorialization in line with the strategic imperative of redistributing productive forces and social surpluses amongst competitive administrative units and territorial locations (Brenner, 2000). In a similar fashion, Peck (2003: p. 357) refers to national states as 'scale managers' who continue to play a key role in shaping discussions about scalar shifts in regulatory capacity, narratives about change and reform, and disseminating knowledge about experiences elsewhere. This implies that the devolution and regionalization processes instigated by national governments may be conceived as a short-term regulatory fix and reflective of a crisis-laden state. Brenner (2000) observes that the re-territorialization process occurring in Germany has been a response to the perceived weaknesses in the Rhone model of capitalism and thus reflects a centrally induced attempt at crisis management. This, in turn, he argues, results in the creation of unstable, evolving regulatory forms which serve only to widen uneven development and rescale the effects of national economic stagnation and increased fiscal austerity downwards into struggles over the territorial and scalar organization of the state. The functions of the national state are thus rendered more residual, more complex and more splintered and have to embrace a more heterogeneous collection of actors in interlocking, multi-level governance and policy communities as a result (Cerny, 2007).

To add further complexity and scope for instability, regions are themselves subject to multiple spatial and scalar selectivities. In other words, whilst the conceptualization and practices of governance form at the regional level may be dominated by the economic and the perceived importance of the regional scale for the successful fulfilment of coordinated economic support, a multi-faceted range of non-economic ideas also underlies the shaping of imaginaries and institutions around the region. As Lagendijk (2007: p. 1202) observes, the privileging of economic discourses in the region 'does not mean that economic development, notably its neoliberal connotation of "competitiveness" is also framed as the primary goal or condition'. Indeed, it is increasingly recognized that economic-competitive rationalities are not the only factors shaping the political geographies of regional governance. This means that there are clear dangers in conflating the re-territorialization of the state with the imperative of competitiveness (Ward and Jonas, 2004). Brenner (2002), for example, takes issue with the fact that New Regionalism and the emergence of regional (and metropolitan or city-regional) governance structures can be understood as stemming entirely from a singular, 'globalist' political movement, and argues, rather, that they are the product of place-specific struggles and the reaction of various social and environmental groups *within* regions to restructuring processes. Regions are thus more heterogeneous and pluralistic as a result.

This may result in competitiveness having to strategically accommodate a range of non-economic, social, cultural, ecological or territorial objectives. Thus, the economic imperatives facing regions may perhaps be framed in relation to symbols and stories around goals such as social cohesion, sustainable development, participatory governance and community development. Indeed, this finds resonance with Dannestam (2008), who demonstrates the tendency for welfare and redistributional policies in Swedish regions to be redefined and legitimized in support of the growth orientation of competitiveness policies. Similarly, in a study of changing policies in relation to housing and the wider support of those workers deemed 'key' to the competitiveness of place, Raco (2008) has demonstrated how welfare and social policies in England are being reconfigured to support economic competitiveness and the pursuit of growth in the name of developing sustainable communities. Interestingly, he draws attention to the 'reactive' way in which these key-worker support programmes have been forged in response to a range of pressures including employer criticisms about skills shortages, political pressure from development agencies in fast-growth regions, and general complaints about the ability of the planning system to support and sustain the means of social consumption. Inasmuch as the political geographies of regional governance are thus more pluralistic and complex than might be supposed from a simple reframing or rescaling of neoliberal ideas, then the goal of competitiveness may also be more fundamentally contested.

Furthermore, the precise spatial scale at which these struggles and contests may emerge is necessarily contingent. Politics is always territorial but the scale of territoriality is a contingent matter (Ward and Jonas, 2004). This inevitably means that achieving some form of coherence is a major challenge for regions. Regions are fragments in a much wider and complex space of policy making and action which is itself filled with multiple ideas and sometime spurious or partial knowledge (Lagendijk, 2007). The salient question thus becomes how meaningful is the notion of a collective regional interest, and, furthermore, how are the priorities of competitive strategies actually constructed and by whom (Knapp and Schmitt, 2003). Knapp and Schmitt (2003) argue that the mobilization of a representative coalition of diverse local interests in support of collective competitive action can never be taken for granted and that conflicts of interest will occur if territorially orientated strategies are to be pursued. Furthermore, they observe that where competitive policies do emerge, they are likely to depend upon a small core of influential stakeholders with particular interests (typically favouring larger firms, international business, high technology sectors and particular groups of workers). This resonates with Lovering's (2003) identification of a 'regional service class' whose interests prevail and are skewed towards meeting the demands of the external competitive environment. The consequent political challenge is to explore both how the benefits of competitive success can be extended more widely across different groups and interests, and how possible trade-offs and tensions with other important objectives such as employment quality, local services and employment conditions can be managed.

Lagendijk (2007: p. 1205) concludes that it is the task of regional institutions of governance to cope with these various centrifugal forces and instabilities such that 'together with local business, state and community organizations, they have to weave an image of coherence, functionality and identity through a myriad of programmatic activities'. This is clearly no small task and demands 'the successful realization of specific regional projects that unite diverse social actors around a distinct line of action in the regional interest' (Hudson, 2007: p. 1154). Moreover, even if this coherence can be achieved, there is no guarantee that its effects will necessarily be predictable, given the 'inability to anticipate the emergent properties of practices' (p. 1154).

Regional governments with their variously imparted vested powers and responsibilities are arguably in a much stronger position than unelected development agencies or 'quangos' to achieve this coherence. Cerny's (2007) observations about the role of US state governments are pertinent here. These governments can claim only partial loyalty from their inhabitants, and have fairly limited powers over internal economic and social structures and forces. Nevertheless, 'they do – like counties, provinces and regions in other countries – foster a sense of identity and belonging that can be quite strong' (p. 272). More especially, their ability to control development planning and shape key variables relating to hard and soft infrastructure (such as transport, education and training), provides them with significant capacity to influence the provision of immobile factors of capital, 'indeed more than many governments in Third World countries' (p. 272). On a practical level, regional governments may also be better equipped to coordinate different policies and agendas inasmuch as they can provide for effective leadership and a place-based perspective for joining up sectoral policy fields with the capacity to respond in a coherent manner (Turok, 2005).

More fundamentally, regional governments can also take their own strategic action to develop alternative repertoires of regional discourses and development agendas. There is indeed a growing literature highlighting the varying degrees of scope for devolved regional governments to choose policy development pathways which diverge in some way from established national patterns or comparative regional norms, particularly in relation to the UK, which has undergone regional devolution relatively recently and in a somewhat asymmetrical and potentially highly divergent fashion (see Jeffery and Wincott, 2006; Trench, 2007). Elsewhere, there are various instances of US states resisting the Washington doctrine of non-compliance with Kyoto commitments on greenhouse gas reduction, whilst several European regions have at times acted to subvert the prevailing political and institutional division of power and resources (Lagendijk, 2007). Autonomous or semi-autonomous regions thus have the potential to develop 'an empowering capacity to resist the intentions of central government ... for the elaboration – although not necessarily the implementation – of alternative regional projects, indicative of the more general contradictory tendencies that plague state policies' (Hudson, 2007: p. 1154).

This creative capacity is clearly, however, subject to certain material constraints (notably, as considered above, in relation to autonomy and budgetary resources), and may be more limited or fail to emerge in the space created for it, particularly in regions suffering some form of institutional sclerosis or history of economic decline and outward migration of talent (Hudson, 2006). It is also likely to be highly contested and contingent, reflecting the increasingly complex and at times congested governance structures within which regional governments increasingly have to operate. Indeed, running in parallel with the process of devolution is the process of 'filling in' whereby institutions and organizational structures are being reconfigured in many and variable ways within regions (Jones et al., 2005b). This results in the reshaping of old institutions, the emergence of new ones, and changed working cultures and institutional relationships.

This gives rise, in turn, to new forms of sub-national politics. In exploring the consequences of the devolution of limited powers but extensive responsibilities for the English RDAs, Harrison (2008b) has coined the term 'regionally orchestrated centralism' to describe the resulting tendency for authority to be rescaled between regions and sub-regions. The RDAs, he argues, are progressively drawing back powers from sub-regional, area offices and programmes into their own control and effectively disempowering the regions. This directly parallels their relative disempowerment by the national state. This is creating an increasingly tangled, variable and potentially contested subregional institutional landscape (Pearce and Ayres, 2009) which has parallels in other national contexts experiencing similar processes of governance restructuring. At the very least, this challenges the traditional conception of regional politics as sub-national and thus defined in relation to the national state (Dannestam, 2008). It is instead simultaneously sub-international, sub-national, sub-regional and local, as new governance relationships and networks criss-cross an increasingly complex and congested institutional landscape.

All of this implies that a complex, dynamic and contingent set of processes, often politically charged, is involved in the production of regional and sub-national spaces (Harrison, 2008b). However, further work is needed to understand these and how they mould and shape the discourses and strategies adopted within regions as a result. The capacity for regional governments to manage agonistic debates and conflicting interests within regions and thus to act as putative agents in forming key coalitions around the competitiveness discourse is of particular interest and is now explored further with reference to the case of Wales.

The Welsh Assembly Government and competitiveness: from a 'Winning' to a 'Vibrant' Wales

The context: economic imperatives, constrained powers and multiple obligations and interests

The Welsh economy was traditionally dominated by heavy industries such as coal, steel and agriculture. The decline of these industries has largely run its course, but it has left an enduring legacy of an under-performing economy characterized by relatively low rates of economic activity and wages. Indeed, Wales witnessed the long-term decline of its per capita GDP from around 85 per cent of the UK average in the late 1980s to around 80 per cent by the year 2000 (WAG, 2002). The poorest parts of Wales, namely West Wales and the South Wales Valleys, were consequently awarded some £1.2 billion of EU Objective One funding between 2000 and 2006. This programme of support is designed to promote structural adjustment in regions with a GDP per head of less than 75 per cent of the UK average. In Wales, this represented a significant economic development investment, with the programme covering some 63 per cent of the area of Wales and 65 per cent of the population (around 1.9 million people).

Delivering the Objective One programme became one of the first and most significant tasks of the newly established devolved tier of government in Wales. In July 1997, the UK Labour government published its proposals for devolution for Wales, which were endorsed by a referendum of the electorate on 18 September 2007. The UK Parliament subsequently passed the Government of Wales Act which established the National Assembly for Wales and laid out its powers and responsibilities. The name 'Welsh Assembly Government' (hereafter Assembly) was introduced on 1 March 2002 to make a distinction between the actions and decisions of the executive (ministers) and the work of the Assembly Members as a whole.

Significantly, key economic arguments were deployed in support of devolution in Wales. In particular, it was argued that further economic development required that the large number of 'quangos' (quasi-autonomous non-governmental organizations) in the region be democratized. Devolution was also intended to capture the economic dividend widely perceived to ensue from the decentralization of power and the capacity to develop more regionally attuned economic development policies through a distinctively 'Made in Wales' approach (CBI Wales, 2002).

The UK government's agenda for regional devolution was inherently asymmetrical, with varying degrees of power devolved from the centre and distinctive arrangements made for each of the UK's constituent nations and regions. Thus the devolution arrangements for Wales ostensibly sat between the extremes of the Scottish Parliament (which was given primary legislative and limited tax-varying powers) and the English RDAs (given a specific remit to deliver economic development and to remain strictly accountable to central government). The Assembly was given no power over taxation, macroeconomic policy, defence or foreign policy or to pass primary legislation. It was instead given the power to pass secondary legislation and the freedom to allocate the budget within its control. It was also empowered to restructure the quangos.

The most distinctive feature of the Assembly was thus the organization of its executive. The Assembly adopted a cabinet system of executive authority chaired by the Assembly First Minister. However, the local government committee system was retained to act as a counterweight to the cabinet and to fulfil a policy development (as well as a scrutiny) function. Thus, what was created was ostensibly a constitutional hybrid and semi-autonomous government involving a balance of power between the executive committee (cabinet) and the various subject committees of the Assembly. The devolution settlement for Wales was thus one based on compromise and caution, with the Assembly being established with constrained powers and as part of a somewhat 'mix and match' pattern of devolved institutional architecture (Bristow and Blewitt, 2001). Indeed, one of the great paradoxes of Welsh devolution is that, whilst more was expected of it in economic development terms, it was given fewer constitutional powers than the Scottish Parliament (Morgan, 2007).

In the early stages of the devolution process in Wales, traditional economic interests appeared to dominate. Indeed, the White Paper 'A Voice for Wales' (Secretary of State for Wales, 1997) identified an unequivocal focus on the continued promotion of an economic agenda. It asserted that 'one of the Assembly's most important tasks will be to provide clear leadership and strategic direction to boost the Welsh economy' (para. 2.11), and was littered with references to the need to create a 'favourable business climate' and 'promote economic competitiveness and growth'. The importance of this task was given added strength by the White Paper's description of the new Welsh Development Agency (WDA) (to be formed by the merger of the old WDA, the Development Board for Rural Wales, DBRW, and the Land Authority for Wales), as the region's new 'economic powerhouse'.

This apparent privileging of economic interests, coupled with the increasingly pervasive expectation surrounding devolution's potential to create the political space and opportunity to shift long-established agendas, provoked a substantial reaction from environmental interest groups in the region. These interest groups, which were previously somewhat peripheral in policy terms, subsequently came together to form the Sustainable Development Charter Group to champion an alternative development agenda. The group was established to debate the environmental policy implications of the devolution process and to provide a coordinated sustainable development lobby of Parliament during the passage of the Government of Wales Bill. It represented a collection of over 25 environmental NGOs, research centres and key government agencies in Wales (such as the Countryside Council for Wales and the Environment Agency). Through intensive lobbying it succeeded in securing the insertion of a clause (section 121) in the Government of Wales Act requiring the Assembly to 'make a scheme setting out how it proposed, in the exercise of its functions, to promote sustainable development'. This clause has two critical and defining features. Firstly, it is a statutory obligation which the Assembly cannot delegate and which has to be approved and debated in plenary sessions. Secondly, it can only be revised and maintained through an inclusive and open process of deliberation and policy making (Bishop and Flynn, 1999). It thus represents a highly unusual constitutional obligation.

A Winning Wales

The Assembly's sustainable development obligation has been upheld as an important step forward in the 'greening of government' and was accompanied by some positive practical steps in building environmental concerns into policy deliberations with, for example, the creation of a Sustainable Development Unit (SDU) in the Assembly Executive (Bishop and Flynn, 1999). In addition, the inclusion of the sustainable development clause and the manner in which it was secured propelled the Assembly's development of a more inclusive, collaborative style of government and governance. Indeed, the emphasis upon inclusive policy making was also enshrined in the Government of Wales Act, which stipulated that the newly devolved tier of government should ensure that its functions conformed to the principle of equal opportunities for all. More specifically, this was buttressed by a unique statutory obligation on the part of the Assembly to consult with the business sector, local government and voluntary organizations in the pursuit of its strategic policy development. Thus, wider interests were explicitly enrolled into the Assembly's policy making and deliberation processes. This was subsequently formalized with the establishment of formal partnership councils between the Assembly and with local government, the business community and the voluntary sector. These partnerships, which have been described as the 'golden threads' of Welsh devolution are unique to Wales (Chaney and Fevre, 2001). Their creation, coupled with both the UK government's continuing enthusiasm for collaborative governance and the EU's requirement for partnership working in the delivery of Structural Fund programmes, has provided a very powerful impetus for the proliferation and growth of multi-sectoral partnerships across all areas of public policy in Wales (see Bristow et al., 2008).

However, these statutory commitments to sustainable development and inclusive governance were not in themselves enough in the early years of devolution in Wales to temper the dominance of the competitiveness imperative. The temptation to prioritize economic goals and imperatives initially proved too strong to resist, such that sustainable development remained a marginal discourse. Instead, the pursuit of competitiveness was the ultimate goal. Thus, for instance, the new SDU unit was situated firmly within the Environment division of the Assembly government rather than being positioned as a more cross-cutting, horizontal unit with a potentially more mainstream function and role. Moreover, the environmental lobby failed in its attempt to secure the establishment of a Sustainable Development Programme Committee to cut across the Assembly's other subject committees, largely because of the strength of the lobbying power of economic interests (Bishop and Flynn, 1999).

More fundamentally, the first national economic development strategy devised by the Welsh Assembly Government was overwhelmingly permeated with the competitiveness discourse, with its title – 'A Winning Wales' – leaving no room for doubt about the dominant ethos at its core (WAG, 2002). The competitiveness imperative was also clearly evidenced by the strategy's central focus on raising the level of GDP in Wales *relative* to the UK average, a clear indication of the overriding emphasis being placed on raising competitive performance as compared to other UK regions. As the strategy put it, 'success would mean Welsh GDP per person rising from 80% to 90% of the UK average over the next decade – with the ultimate aim of achieving parity. This is the main goal of our economic policies ... ' (WAG, 2002: p. 20). The strategy also made explicit reference to the OECD's identification of the four main drivers of economic growth and competitiveness, i.e. innovation, enterprise, people, and the application of information, communications and other technologies. Indeed, these four factors were translated into the key priority areas for action in Wales.

More specifically therefore, the strategy asserted firstly that 'innovation drives both competitiveness and sustainable development' (WAG, 2002: p. 9). In seeking to raise the proportion of business expenditure on R&D, the strategy thus emphasized the importance of strengthening the innovation culture in Wales and of building on existing business clusters and technology fora, especially in fields such as opto-electronics, information technologies materials and biosciences. This was subsequently fleshed out in the publication in 2003 of the Assembly's Innovation Action Plan, which was explicitly designed to 'deliver on the commitment in "A Winning Wales" to make Wales more competitive within the global economy' (WAG, 2003: p. 4). As such, it focused exclusively on expounding the actions deemed necessary to promote competitiveness and the knowledge-based economy through business innovation (with explicit reference made to Porter's work) and, in particular, expanded university–business links. A key policy instrument outlined was that of the Technium, which, in line with the UK government's emphasis on supporting building incubators as 'seed crystals' for clusters (Cooke, 2003), was designed to offer hosting facilities for university-based business start-ups and other high-tech businesses. Some 20 Techniums were thus planned as key focal points for innovation across Wales, in conjunction with other initiatives such as the Knowledge Exploitation Fund (KEF), which was designed to encourage a step-change in knowledge commercialization within the academic base in the region.

Secondly, 'A Winning Wales' set the growth of entrepreneurship as a key priority, with this to be measured by a growth in the stock of VAT-registered businesses in the region, and to be delivered through an Entrepreneurship Action Plan aimed at supporting business survival and growth rates. This marked the end of the WDA's love affair with inward investment, which had seemingly dried up and which was part of the increasingly defunct and outdated, smokestack-chasing approach to regional policy (see Chapter 2; also Cooke, 2003). Thirdly, 'A Winning Wales' also identified skills and learning enhancement as being critical to the economic development effort, with targets set for both reducing the numbers of people with no qualifications and increasing the numbers of those with higher-level qualifications. Finally, the strategy stated that 'Wales urgently needs a globally competitive communications system' (WAG, 2002: p. 12), and thus set targets for the uptake of ICTs in business, the growth of e-commerce and the spread of a broadband infrastructure.

Thus 'A Winning Wales' was clearly in tune with the EU's Lisbon agenda and the UK government's emphasis on competitiveness and the knowledge-based economy. This is understandable, given both the size of the prosperity gap between Wales and the rest of the UK and the strong economic *raison d'être* underpinning the Assembly's very existence. Furthermore, the size and scale of the EU's Objective One programme in the region at the time provided a very powerful impetus to the EU's competitiveness agenda and unerringly skewed spending priorities in Wales towards economic development (Cooke and Clifton, 2005). However, this made it difficult for sustainable development to be anything other than a peripheral, incommensurable and weak concept, overtly focused on environmental conservation, which thus posed no threat to the vision of competitiveness and its associated pursuit of relative growth. As Mainwaring et al. (2006: p. 688) put it, 'the view that WAG's obligations are consistent with rapid GDP growth requires a conception of sustainability that is so broad that it effectively marginalizes the interests of future generations, within Wales and without, a view that is perilously close to sophistry'.

Sustainable development was ostensibly an add-on to the competitiveness imperative at this time, rather than being in any sense integrated with it. This is indeed implicit in the description of the strategy made in its preface by Rhodri Morgan (the First Minister in the Assembly), who referred to it as 'the WAG's strategy for transforming the economy of Wales, whilst promoting sustainable development' (WAG, 2002). Whilst some attempts were made to link the discourses together in the document with statements such as 'economic growth is not sustainable where the interests of the environment and our established communities are disregarded' (p. 2) and 'the outstanding natural environment of Wales and our comprehensive higher and further education networks are invaluable assets for Welsh business' (p. 2), policy actions were overwhelmingly couched in terms of the economic. The sustainable development commitments in the strategy were fairly narrow and confined to discrete actions around reducing waste and promoting the quality of the natural environment for tourism. Not surprisingly as a result, the Assembly's early progress in respect of its sustainable development obligations was criticized as being of marginal significance, with considerable variability evident in the commitment to sustainability across the Assembly-sponsored public bodies, and with successes regarded as being limited to improvements in 'talking the talk' of sustainability rather than tangible actions (or 'walking the walk') (CAG Consultants, 2003; Bishop and Flynn, 2005).

Wales: A Vibrant Economy

Tellingly, the Welsh Assembly Government revisited its economic development strategy in 2006, with a view to ensuring that its 'economic development activities dovetail ever more closely with the action [being taken] on the social and environmental dimensions of sustainable development, both nationally and in local areas throughout Wales' (WAG, 2006: p. 1). Its new strategy, revealingly entitled 'Wales: A Vibrant Economy' (WAVE), set out a new vision for the region, one based on the delivery of 'strong and sustainable economic growth by providing opportunities for all' (p. 6). Thus, the Assembly's economic agenda was increasingly viewed as inextricably linked with its broader strategic commitments to social justice, balanced spatial development and, most notably, sustainable development and an improved quality of life. Indeed, the Assembly's environmental objectives and obligations appeared to have moved much further up the list of policy priorities, so much so that its economic development objectives (which remained focused on 'helping business to become more competitive by supporting ... entrepreneurship, innovation, investment and trade' (p. 5)) were redefined in relation to, and in support of, its green agenda. Critically, it identified one of its key actions as 'ensuring that all economic programmes and policies support sustainable development, in particular by encouraging clean energy generation and resource efficiency' (p. 5). In so doing it established that sustainable development was now regarded as 'central to the approach' (p. 9), and placed greater emphasis on green business development through, for example, additional support for a low carbon economy, as well as the development of renewable energy and biomass development. Furthermore, it developed a strategy for sustainable public sector procurement designed to encourage greater emphasis on spending public money locally and with respect to social, economic and environmental objectives through, for example, local sourcing of ingredients for school food (Morgan and Morley, 2006).

This change of emphasis was affirmed by the downgrading of GDP (and GVA) targets in the overall strategy, which were criticized as failing 'to capture the full impact of an economy on quality of life' (p. 67). Thus, in relation to its previously stated ambition to raise the quality of life in Wales to match that of the UK as a whole within a generation, the Assembly Government indicated that it had realigned the focus of its efforts towards increasing employment and raising value-added per job and earnings, 'as these relate directly to improvements in the quality of life for people in Wales' (p. 67). Furthermore, in line with this more pluralistic development agenda, it broadened the set of indicators to be used to assess overall performance with a view to providing 'a more balanced and appropriate way of assessing progress in the Welsh economy' (p. 69). Thus, for the first time, sustainable development was reflected in the Assembly's assessment of progress in delivering the economic agenda through the establishment of a key role for environmental 'satellite' accounts, as well as for more holistic progress metrics such as the ecological footprint (which seeks to measure the environmental resource costs of existing lifestyles and patterns of consumption), and the Index of Sustainable Economic Welfare (ISEW), which seeks to adjust GDP for the social and environmental costs of growth. In short, whilst the competitiveness imperative remained strong, it had been somewhat rebalanced with a much greater emphasis on sustainability and quality of life, to effect a more hybrid and locally contextualized development strategy and discourse.

There are a number of explanations for this change. Of particular importance is the weak devolution settlement which confronted Wales after 1997. Indeed, there appears to have been a progressive acknowledgement within the Assembly Government itself of the inherent difficulties it faced in meeting its competitiveness agenda, particularly in the absence of comprehensive legislative powers, resource-raising capacities and responsibilities. The potential for implementation failure was indeed summarily observed by Michael Porter himself, who criticized 'A Winning Wales' for constituting little more than a 'wish list', for having no obvious mechanisms for delivering success and for paying insufficient attention to the provision of proactive support for improvements in the Welsh business environment and the development of clusters (Porter, 2002).

Cooke and Clifton (2005) argue that the limited powers granted to the Welsh Assembly Government, coupled with its over-interpretation of EU requirements for inclusive governance, resulted in the early years of devolution in Wales being characterized by a 'precautionary' and ostensibly risk-averse approach to economic governance (see also Bristow et al., 2008). The effectiveness of interventions such as the Entrepreneurship Action Plan, Knowledge Exploitation Fund and Technium concept have proved to be limited to date by their over-ambitious estimates of the potential for university–business spin-outs in the region, coupled with their tendency to replicate old incubation approaches that failed to provide sufficient management assistance and move beyond the simple provision of property leasing space.

The Assembly's most significant actions were more typically confined to the reorganization of the administrative apparatus and the creation of a complex and somewhat cumbersome public sector bureaucracy for the delivery of the Objective One programme – effectively the areas where it clearly had the powers to act. Indeed, by 2004 it was estimated that some 1,700 civil servants in Wales costing £36.2 million per annum were managing economic affairs in the Welsh Assembly Government, a 25 per cent increase over 3 years (see Cooke and Clifton, 2005). Moreover, on 14 July 2004 the Assembly took what Cooke and Clifton (2005) regard as the ultimate precautionary measure of announcing that the three main economic development quangos (the WDA, Education and Learning Wales, ELWa, and the Wales Tourist Board, WTB), would be terminated by April 2006 and absorbed into the Assembly's economic development and skills departments. This so-called 'bonfire of the quangos' signalled an intent to bring key aspects of the decision-making process into the political arena and electoral process (North et al., 2007) and was indeed justified on the basis of improved ministerial accountability and more efficient service delivery, rather than on the basis of its ability to create enhanced competitive performance (WAG, 2006). As such it represents a clear instance of the re-statization of the organizations of economic development in Wales (Jones et al., 2005b).

Perhaps not surprisingly, the combined result of the Assembly's efforts to date has been little more than glacial progress in meeting its economic development objectives, and specifically improvements in competitiveness through relative growth in GDP per head (see Bristow et al., 2007). By 2006 and the publication of WAVE, the Assembly was, however, mature enough to acknowledge its own limitations in respect of economic governance and to assert with greater confidence the possibilities for a different strategic focus. In emphasizing the importance of both a stable macroeconomic environment and competitive and flexible capital, labour and product markets for business competitiveness, the devolved body was confident enough to admit that 'the Assembly Government has relatively few direct responsibilities in these areas' (WAG, 2006: p. 14). It was also then in a position to define more clearly where its scope for most effective intervention lay (namely, in improving business infrastructure through its influence over transport, land and property development), as well as to assert its progressive success in securing greater responsibilities and in moving up the steep learning curve associated with the early period of devolution: 'the Assembly Government's growing range of devolved powers, short decision chains, close partnerships, local knowledge and willingness to engage will help in building an ever stronger competitive advantage for Wales' (p. 17). Indeed, such were the limitations of the original devolution settlement that pressures for greater parity with Scotland quickly gathered momentum, resulting in the new Government of Wales Act 2006 and its identification of a road map for greater legislative powers in Wales.

This highlights the often-made claim that devolution is a process, not an event, and it is clear that 'devolution has been a highly protean political project in Wales in recent years' (Morgan, 2007: p. 1240). Thus, as well as experiencing progressive evolution in its confidence, capacities and powers since its inception, the devolved Assembly Government has also presided over, and helped to create the space for, the emergence of a reinvigorated and more diverse sub-national politics. The changed nature of politics in Wales is best evidenced by the signing of the One Wales document, which constitutes the agreement reached in June 2007 between the Labour party and Plaid Cymru (the Welsh nationalist party) to share power and deliver the Assembly's policy programme for its third term. This document has contributed significantly to the enhanced emphasis on sustainable development within Wales, with a key commitment being the establishment of a Climate Change Commission for Wales capable of assisting with the development of new green policies and the establishment of targets for cutting greenhouse gas emissions.

The establishment of the Assembly, coupled with its particular predisposition towards inclusion of diverse interests, has certainly therefore created new spaces for inclusion and a new system-openness within which previously excluded voices and alternative agendas can be heard. Opportunities for engagement in decision-making processes have been significantly improved in Wales, particularly for voluntary sector organizations which place a high value on their enhanced access to key personnel and policy networks and the formal recognition of their 'seat at the table' provided by the statutory underpinning of the inclusion principle (Bristow et al., 2008). Wales is thus characterized by a more consultative and deliberative politics as a result, although the extent to which there is scope for radically different development agendas to emerge is debatable.

On the one hand, the new spaces of engagement and inclusion have allowed lobby groups and political interests around sustainable development, social inclusion and balanced development to gather momentum. For example, the growing clamour within Wales to address sub-regional disparities and, in particular, the high concentrations of disadvantage in the Heads of the Valleys area of South Wales, has helped to propel the publication of the Wales Spatial Plan, with its clear commitment to address variable local development priorities within Wales (WAG, 2004). On the other hand, this has arguably simply encouraged the rescaling of the competitiveness discourse down to sub-regions, notably the Cardiff city-region, which has a strong track record of pursuing boosterist agendas (Hooper and Punter, 2006). Furthermore, creating inclusive structures of governance in a relatively small region with limited policy-making capacity has encouraged negotiation amongst cross-sectoral local elites and a resulting tendency towards consensual, centrist, 'third way' decision making (Bristow et al., 2008). Indeed, one possible further effect of the 'bonfire of the quangos' might be to reduce the scope for dissonance or variation in the policy discourse within Wales. Only time will tell if the competitiveness discourse in Wales will be more significantly challenged as opposed to, as at present, simply being more 'made in Wales'.

Conclusions

Using the CPE framework, this chapter has identified and explored more fully the range of different variables with the potential to shape the transmission of policy ideas around regional competitiveness into tangible interventions and models of governance in practice. In particular, it has focused on exploring the relative importance and agency of devolved regional governments and their capacity to identify, represent and mobilize a coherent, collective vision amid competing interests and agendas, as well as their power to decide on and deliver transformative action. In so doing, this has responded to the call for a more sophisticated understanding of the relationships between processes of state rescaling and the spread of key economic imaginaries such as competitiveness. It has also allowed for greater consideration of the extent to which recontextualization in the discourse of competitiveness across regions leads to its continued dominance through 'requisite variation' or the scope for more obdurate and transformative resistance.

The case of Wales has been used as an example of a region in the throes of establishing and embedding a devolved government and experiencing highly pertinent debates around its strategic imperatives and capacity for action. In Wales, it is clear that the competitiveness imperative was initially uppermost in the minds of those devising its economic development agenda, in spite of a strong legal obligation to deliver on an alternative discourse, that of sustainable development. The region was intent on developing a 'winning' strategy geared towards the ultimate objective of reducing its relative prosperity gap with the UK as a whole. Over time, however, a more locally attuned, hybrid discourse around 'vibrant' and 'sustainable growth' has emerged which seeks, if possible, to more effectively reconcile or balance the pursuit of improved economic competitiveness with the desire for sustainable development and an improved quality of life. The revised economic development strategy interweaves elements of the competitiveness and sustainable development discourses and also incorporates new conceptual elements that have emerged out of the debates articulated within the new, more inclusive governance framework instituted by devolution in Wales. These include a concern with regeneration, social inclusion and, in particular, balanced territorial development across its different sub-regions – the latter perhaps implying that the region as constituted is not a particularly coherent scale for the pursuit of competitiveness. All in all, this provides valuable insights into the hybrid nature of Wales's actually existing economic development strategy which features a robust, albeit dynamic and evolving adherence to the competitiveness ethos.

To date, however, this evolution of strategy appears to represent the recontextualization and restating of competitiveness to fit the changing interests being articulated by the Welsh Assembly Government, rather than a more fundamental and radical resistance to the competitiveness goal. At present, competitiveness thus still seems to subvert other objectives to its aims, such that only requisite variation is in evidence. This reflects fundamentally the constrained capacity of the new body to challenge the discourse and intent of both the national state and supranational governance regimes that continue to effect considerable influence through 'centrally orchestrated regionalism' and the managerialism of key policy programmes. In Wales, this has been reflected in two key developments. The first is the primacy afforded to delivery of a large-scale EU Objective One programme and an associated and immediate focus on creating the requisite procedural and organizational bureaucracy. The second is a precautionary policy approach to economic development based around copying (arguably rather badly) the innovation and supply-side policy approaches adopted elsewhere, and a reshuffling of its own institutional architecture through reorganization of the quangos. Thus, in a manner similar to the experience of other newly devolved institutions around the world, it appears that the early years of the Welsh Assembly Government were blighted by a degree of sclerosis and an introspective focus on the inflation of its own bureaucracy – with inevitable and associated costs (see Rodriguez-Pose and Gill, 2005). This was largely on account of its effective subjugation to the authority and parameters of higher levels of government. The Assembly lacked any real power to influence the key factors shaping business competitiveness, and inevitably took time to garner the requisite knowledge, information and confidence to understand this and to assert itself accordingly. This highlights the ultimate significance of the degree of regional autonomy that exists. The empowerment of regional interests through devolution, particularly where regional capacity for self-determination is limited, does not appear to be sufficient on its own to provide a profound disorientation to the competitiveness imperative, or indeed to produce fundamentally different economic outcomes (see also Hudson, 2006).

Notwithstanding this, the case of Wales has clearly highlighted that the devolutionary process and its accompanying processes of re-statization are highly dynamic and subject to considerable flux, such that they do, over time, help to create the space for the further evolution of processes of variation and resistance. In Wales, the nature and role of the regional government itself is evolving and is subject to continued development in respect of its powers, capacities, responsibilities and confidence. This is in turn being accompanied by, and is itself also helping to stimulate, the process of 'filling-in' whereby the regional institutional landscape is being reconfigured and new working relationships and organizational cultures are being formed as a result. The nature and significance of sub-national politics is also experiencing further development and change as more diverse and potentially dissonant voices are being enrolled into decision-making processes. As a consequence, the regional development agenda is being shaped to a much greater extent than previously by the needs of pluralism and compromise. The pressure is thus growing within Wales to more effectively deliver on its sustainable development obligations, whilst the Assembly, through the One Wales agreement, is pressing the national government for further reviews of its legislative powers and the mechanisms currently used to determine its resources (see, for example, the work of the Independent Commission on Funding and Finance for Wales).[1] There is thus ever increasing scope for agonistic debate both between the region and the national state, and within the region itself.

The outcomes of this are difficult to discern and predict and will principally be determined by political debate and compromise. What is clear is that the current compromise between competitiveness and sustainable development in Wales is likely to be both an uneasy and an unstable one, not least because of the inherent trade-off between the environment and economy (Mainwaring et al., 2006). This, in turn, provides scope for further evolution and change. As Hudson (2006: p. 169) asserts, 'any notion of the region as a unified and coherent political subject pursuing policies (for example, via an elected Assembly) to further a shared regional interest is untenable'. Different interests will collide in the ensuing political struggle, with the inevitability that distributional issues of equity within the region will assume greater importance, and the distinct possibility that alternative conceptions of development may ultimately emerge and flourish.

This example serves to illustrate some interesting tensions which surround the role of regions and territorial governance structures in particular. In part, these structures are constituted and legitimized by the competitiveness imperative. They owe their very existence to its logic and it, in turn, provides some structured coherence for their policies and practices. Yet at the same time they are clearly rich sites for contest between different scales, inasmuch as they are often forced to mediate between the interests of local, national and supranational actors and institutions, whilst also providing a focus for the mobilization and collision of different social and environmental interests and groups. The inevitability of regional state spaces as spaces of competitiveness may thus have been somewhat overstated.

Part III
Moving beyond competitiveness

6   The limits to competitiveness

Introduction

The preceding chapters in this volume have demonstrated that whilst competitiveness is a broad, nebulous and somewhat chaotic discourse, it is an
innately powerful one that commands widespread support. Indeed, it has
become a hegemonic discourse shaping economic development thinking,
policy and strategic action in regions around the globe. Its position is such
that whilst it can be moulded and shaped to fit different circumstances as it
spreads across different regions with their different constituent interests and
powers of agency, it nonetheless appears to retain a seemingly unshakeable
hold over policy thinking and practice. In fact, the ability of the discourse to
'mutate' in this way perhaps makes it more robust and embedded in regional
thinking and policy practice. In turn, the particular conception of competitiveness that dominates provides a clear focus to policy intervention and
action. The pre-eminent conception of regional competitiveness as equivalent
to 'attractiveness', or the capacity of the region to compete with other places
for mobile capital, leads to a strategic emphasis on the ability of the region to
attract and retain innovative firms, skilled labour, mobile investment and
central and supranational government subsidies and funds, and an overriding
focus on the pursuit and measurement of their success in doing so relative to
other places or 'rivals'.

There is, however, growing awareness of the shortcomings of competitiveness
thinking and its implications for strategic policy choices and outcomes in
practice. Krugman (1994; 1997a) has famously derided place competitiveness
as a 'dangerous obsession'. This, in part, reflects his concerns regarding the
validity and relevance of the concept itself in relation to cities and regions for
which, unlike for business, poor economic performance has no bottom line –
places do not ultimately go out of business. However, Krugman also points to
the dangers of the corollary of competitiveness in terms of the nature and form
of policy intervention which ultimately ensues, which he regards as presenting
unnecessary and ineffectual 'meddling' by governments in the concerns of
business. Krugman's belief is that regions initially develop, grow and prosper as
a consequence of particular path-dependent processes, citing the Silicon Valley

cluster in California as owing its existence to 'small and historical accidents that, occurring at the right time, set in motion a cumulative process of self-reinforcing growth' (Krugman, 1997b: p. 237).

Following his lead, other critiques have emerged, with Kitson et al. (2004: p. 997) asserting that 'it is at best misleading and at worst positively dangerous to view regions and cities as competing over market shares, as if they are in some sort of global race in which there are only "winners" and "losers" '. Similarly, Unwin (2006) criticizes the obsession with competitiveness for obscuring the broader elements impacting on city and regional economic performance, whilst Malecki (2004) warns that the dynamics of competition between places are necessarily fraught with negative rather than with positive connotations (see also Peck and Tickell, 1994 who refer to 'jungle law' breaking out).

The purpose of this chapter is to consolidate and extend the existing debate on the limits to regional competitiveness. In particular, it aims to examine in detail the limitations to competitiveness thinking, specifically in relation to its implications for economic development and the nature of policy outcomes in practice. It tells the tale of competitiveness as exemplifying a theory led by policy approach, and examines the limiting implications of this for policy. The focus is on the role of competitiveness thinking in narrowing policy approaches and foreclosing wider analysis and understanding about how regions develop, grow and prosper. This, it is argued, has significant potential to constrain the development of a broader, more radical regional development agenda and politics. Thus, in identifying the need to rethink the merits of the competitiveness imperative for regions, this chapter paves the way for Chapter 7 to consider the potential scope for new approaches to regional development and policy to emerge and flourish in the future.

The persistence of regional inequalities

It is an open question whether and how specific competitiveness policies and strategies impact on economic performance across regions. There is plentiful evidence on which to draw and a growing evaluation industry around different regional development programme and policy interventions, particularly in relation to key competitiveness policy tools and instruments such as clusters, innovation policy and university–business links, science parks and technopoles. The findings from such studies suggest that there is good reason to think that policy can make a significant difference to regional competitive performance, and yet at the same time it is very hard to know exactly what the right policy is (Martin, 2005). The results of such studies will inevitably be variable and case specific and, as such, it is not the intention here to attempt the impossible and to summarize these diverse experiences and identify generic lessons about what works in what instances and with what effects. Instead, the focus here is on the more fundamental questions around the capacity for strategic approaches, under the rubric of competitiveness, to produce positive and sustainable outcomes for regions.

A cynical view is that competitiveness strategies do little more than legitimize high-profile, boosterist developments aimed at attracting events, investment, property developers and people. Such strategies, it is argued, legitimize the diversion of official attention amd public resources and lead to a particular set of iconic projects deemed to lead to 'improved competitiveness'. As such, they serve as a means whereby development is sold to relevant citizens, and they are more an accompaniment to change in cities and regions than a driver of it. Instead, overall patterns of spatial economic change and employment growth are driven more by other, broader influences, namely public spending on collective services, service sectors related to private consumption and the effects of the global boom in property markets (Bristow and Lovering, 2006).

What is clear is that those parts of the world that have pursued regional competitiveness with vigour are still blighted by regional divides between their rich and poor regions. In short, the pursuit of competitiveness appears to reinforce and reproduce uneven development between countries and regions (Birch and Mykhnenko, 2009). From the beginning of the 1980s, regional inequalities within the EU have increased. Per capita disposable income inequalities have significantly widened within each country and within each region. The most recent report on regional economic disparities in the EU, for example, highlights the continuing large development gap between European regions, and particularly between old and new member states. The top 10 regions of the EU in terms of GDP per head are all located in the EU15 and are often capital city regions. At the other end of the spectrum, several regions in Bulgaria and Romania have levels of GDP per head below 30 per cent of the EU27 average (CEC, 2008). As such, there is a very real danger of the emergence of an 'archipelago Europe' within the context of a 'global archipelago', by which is meant a society 'characterized on a spatial level by a strong concentration of techno-scientific, financial, economic and, above all, cultural-political decision power into a restricted number of "islands" of wealth and innovation, surrounded by a sea of "peripheries"' (Petrella, 2000: p. 70).

Similarly, the UK is still characterized by a prominent 'north–south' divide in regional fortunes, the obdurate nature of which provides a rather depressing if somewhat familiar look to the nation's socio-economic landscape (Evans and Pentecost, 1998; Fothergill, 2005; Morgan, 2006). These disparities, particularly between the English regions, are subject to increasing political criticism and debate, with some critics making links between poor economic outcomes and the policy and institutional approaches directly associated with competitiveness. For example, in a recent report focusing on the performance of the English Regional Development Agencies (RDAs), the TaxPayers' Alliance (an independent campaign group) has argued that, apart from London and the South East, England's regions grew faster in the seven years *before* RDAs were introduced than in the seven years after, in both per head and total output terms. Furthermore, the rate of business creation has not significantly increased since 1999 and the relative contribution of the seven regions outside London and the South East to England's economic output has dropped from 64 per cent in 1992 to 52 per cent in 2006. In short, the gap between the richest and poorest regions in England has not diminished (TaxPayers' Alliance, 2008). Regional disparities across the US, Canada and China regularly attract similar political and media attention.

This is indicative of some form of policy failure and raises interesting questions about whether the continued evidence of lagging regions represents the age-old problem of implementation failure, or reflects more fundamental deficiencies in respect of competitiveness strategies and their capacity for improving economic outcomes. There is certainly some evidence which shows that the growing divergence between successful and lagging regions in the EU is reflective of the divergence between innovation-prone regions, where there is strong policy support for innovative firms and innovation infrastructure, and innovation-averse regions, where relevant policy support is much less developed or backward (Rodriguez-Pose, 1999). Similarly, European integration processes and policy approaches geared to competitiveness and growth have been found to work better in some places than in others and thus to widen disparities through their impact in encouraging the agglomeration of economic activities in already successful cities and regions (Petrakos et al., 2005; Geppart and Stephan, 2008).

However, it could instead be that the pursuit of competitiveness is such that insufficient policy attention is being paid to providing targeted support to weaker regions or to focusing on cohesion and the reduction in regional inequalities. Indeed, as already mentioned, implicit in the underlying rationale of regional competitiveness is the assertion that 'everyone can be a winner'. In this context, a growing emphasis on regional competitiveness, particularly through endogenous growth and devolved forms of governance, has been accompanied by a tendency to 'treat unequal regions equally' (Morgan, 2006) and thus dispense with, or at least significantly downgrade, more redistributive, regional policies with a stronger focus on territorial justice or equity. The critical point is that whatever the nature of the failure, there are clear and persistent regional winners and losers from the competitiveness game.

The limits to competitiveness (1): how to implement

Understanding why the winners win and the losers lose requires closer interrogation of the capacity for regional competitiveness strategies to be pursued and implemented effectively across a wide range of regions. Perhaps not surprisingly, given the rather chaotic nature of the concept of competitiveness itself, there are fundamental questions around its utility as a basis for coherent and effective strategy making and policy design in regions. First and foremost, and as Chapter 1 has demonstrated, there is no coherent or agreed theoretical framework for conceptualizing regional competitiveness and thus no ready-made template or blueprint for regions seeking to devise competitiveness-oriented policy interventions to follow. All the different theoretical models that have been alluded to as potential frameworks for understanding regional competitiveness have their limitations, and propose different drivers of competitive performance and outcomes (Kitson et al., 2004). Thus, whilst there may be convergence around certain popular models such as Porter's diamond, there are enough theoretical variants shaping the discourse to make for a diverse range of definitions of competitiveness to deploy in practice, and a confusing menu of possible drivers of competitive performance to prioritize. In this regard, the discourse of regional competitiveness constitutes a classic tale of 'theory led by policy' (Lovering, 1999), whereby commentators have found themselves engaged in *ex post* rationalization of a term that has already become embedded in common policy parlance.

Second, even if regions were able to successfully replicate the Porter model or some such approach, there are no cast-iron guarantees that success will follow. Some regions will necessarily fail 'because their best attempts at developing a competitive advantage may not be good enough' (Collins, 2007: p. 78). In other words, there will always be instances of implementation failure. These are arguably magnified in respect of many of the standard policy prescriptions that have come to define the competitiveness discourse, because they are poor travellers from the successful or exemplar regions from which they originate. History and geography will have a major impact on the relevance and utility of particular drivers and their impact in particular regions, such that there is unlikely to be an effective 'one size fits all' regional competitiveness strategy (Kitson et al., 2004; Bristow, 2005). This is especially pertinent when one remembers that many of the key ingredients for economic success are deemed to be endogenous variables, locally shaped and attuned to particular regional circumstances.

Particular misgivings have been raised regarding the unrealistic aspirations of many regions to develop successful high-tech cluster strategies and policies. These have been roundly criticized for their reliance upon off-the-shelf blueprints which fail to acknowledge that clusters are often context-specific socioeconomic configurations of concentration and collaboration which typically represent the exception rather than the rule in examples of regional economic success, and tend to emerge organically without the benefit of policy guidance or intervention (Hospers, 2006; Burfitt and MacNeill, 2008).

A good example, widely replicated in many locations, is that of Silicon Valley in California. This has resulted in the replication of a high-technology, electronics and software nexus in other 'silicon' places such as Silicon Glen in Scotland. In practice of course, there were and are substantive differences between California and Scotland that resulted in quite different characters as between the electronics industries of the two places. Whereas Silicon Valley was nurtured by huge government spending on military and aerospace and enjoyed the support of a strong venture capital presence that proliferated spin-off businesses, Scotland clearly lacked these attributes. As a consequence, Silicon Glen lacked the self-regenerative growth capacity or resilience of Silicon Valley and largely became home to low value-added electronic assembly operations vulnerable to cost competition from other locations around the world (Bristow and Wells, 2005).

Similar problems afflict the other policy tools that have mushroomed under the competitiveness agenda, notably technopoles, science and technology parks and enterprise zones. In all these cases, a discrete and small spatial area has been designated within which policy unfolded. The spatial logic of location concentration has not always been evident: the businesses that occupied enterprise zones were assumed to need no relationship with other businesses in the zone, for example. In the case of science parks and technology parks, in contrast, there was often an implicit notion that mutual interlinkages would reinforce the competitiveness of the individual companies and hence of the locality itself. The critical point is that the rationale for this different thinking lies purely within the attempt to replicate what seems to have worked elsewhere, rather than any clear underlying logic as to what will work best for particular firms in particular places.

In short, competitiveness policy tools and their strategies typically lack sensitivity to critical issues of context and place. As a result they fail to serve as useful guides for the implementation of economic development strategies. As Markey et al. (2008) observe in relation to north-western British Columbia in Canada, this means that policy instruments are applied which are often more associated with large, urbanized metropolitan environments. This lack of tailoring to context 'leaves development dialogue trapped in the abstract, where reports create false expectations, and where regions may be led towards ill-suited programme interventions based on passing policy or development fads' (Markey et al., 2008: p. 342). They go on to state that 'if economic strategy reports are not contextually specific, they cannot support a local strategy' (p. 342).

This lack of sensitivity to place and context is not surprising. The dominant thinking on regional competitiveness asserts that regions are clearly defined, internally coherent atomistic and bounded spatial entities with quantifiable attributes that are in their exclusive possession, for each of which a desirable competitive advantage can be identified. Thus, they are assumed to compete on a level playing field, in directly commensurable terms and in a manner directly equivalent to firms, with fortune favouring the entrepreneurial. This, however, fails to respect the very different economic and political structures that shape regions and that make them highly unique social aggregations rather than equally comparable economic ones.

Similarly, neither is it entirely clear whether the 'region' is always necessarily the appropriate spatial scale for policy intervention. New Regionalism has come under increasing attack for its tendency to take the fundamental concept of 'the region' for granted, thereby displaying a kind of 'spatial fetishism' that tends to elide intra-regional divisions and tensions (MacKinnon et al., 2002).

It is not always clear, in fact, how the abstract notion of the 'region' used in competitiveness discourse relates to the actual regions in which people and firms reside and to the spatial delimitations and functional areas which may influence competitive advantage (Lovering, 1999). Indeed, the spatial specificities that provide the impetus for economic development and firm competitiveness may be highly localized, or a complex mix of local, regional, national and even global influences. That said, the competitiveness discourse offers little in the way of guidance as to whether and how far policy should concentrate on particular localities within regions with the greatest potential for competitive success through the development of, for example, localized clusters or broader agglomerations of competitive firms. Neither does it afford any insights into the likelihood that the benefits of such an approach will filter out into other parts of the regional economy more generally (Kitson et al., 2004). The result is inevitably a rather broad-brush approach to competitiveness strategy and policy. Indeed, in most instances 'policies are pursued on the basis of predefined administrative or political areas that may have little meaning as economically functioning units, and from which policy effects may "leak out" into other regions' (ibid: p. 997).

The limits to competitiveness (2): constrained policy action

A further set of problems derives from the fundamental axioms which lie at the heart of the dominant thinking on regional competitiveness. These collectively serve to limit the range of policy options explored within regions and help to perpetuate the powerful and pervasive logic that there is seemingly no alternative to the competitiveness agenda and its particular brand of policy approaches.

The first of these concerns the hypermobility of global labour and capital. Once capital is reconstituted at a global level, 'the often instantaneous movement of people, ideas, information and capital across borders means that decisions are swayed by the threat that needed resources [or economic activity] will go elsewhere' (Ohmae, 1995: p. 119). However, the uncertainty that surrounds what causes certain forms of capital to flee or shy away from particular territories is sufficient for states to rationalize that self-discipline in relation to the 'global logic' of market forces is the most feasible or indeed only policy option (Fougner, 2006). This creates an innate imperative for regions to compete in order to provide the appropriate living and working environment to attract footloose firms, investment and highly skilled and creative labour.

The inherent links between the discourses of competitiveness and globalization are thus very clear. For decades, the approach to spatial economic development policy has been to enhance local and regional competitiveness at the same time as the scale of spatial competition has inexorably widened. In general terms, regional and subsequently national economies have been less insular and more integrated into the emergent global economic system. Through a combination of internal political changes and international agreements, such as those negotiated by the World Trade Organization, more countries have been drawn into the market economy framework. Formerly isolationist countries, such as those comprising the former socialist economies of Central and Eastern Europe, as well as China and India, are continuing to follow a process of 'liberalization' to attract investment, find markets for their products and services and achieve economic growth to deliver material prosperity.

In crude terms, if we consider the question of spatial economic development policy in the UK over the last 50 years, the trends are evident. In the era immediately after the Second World War, notwithstanding the reconstruction efforts due to war damage mainly in London and surrounding areas, the main focus of policy was one of rebalancing economic growth across the various regions. In particular, policy sought to attract and retain investment into the 'peripheral' and disadvantaged regions such as Wales and the North East of England. In this respect, regions such as Wales were perceived as being in competition with other peripheral UK regions, particularly in terms of attracting inward investment. The process of European integration somewhat changed this perspective. The stages that culminated in the establishment of the EU and the development of policies designed to redress spatial-structural disparities in the pan-European economy inevitably meant that inter-regional competition became played out on a larger, European stage. The perception then was that Wales was competing with Alsace, or with regions in southern Italy and Portugal. The process of European enlargement, coupled with the removal of barriers to the movement of goods and capital in other countries, has further widened the potential locational choices of manufacturing and service businesses around the world. As a consequence, regions such as Wales increasingly perceive themselves as being in competition with India and China.

Yet in all probability, global capital is not as hypermobile in reality as states perceive it to be. Whilst routine assembly-line manufacturing operations and call centre firms may be globally footloose, there is a strong body of evidence pointing towards the geographical inertia of many businesses, including those firms in 'new' economy industries often regarded as having the greatest potential to be locationally flexible in the wake of new, distance-shrinking technologies (Morgan, 2004b; Rodriguez-Pose and Crescenzi, 2008). Many firms, especially local service providers, retailers and small businesses, simply never relocate – the costs and upheaval of doing so are too great, and the innate advantages of, and ties and preferences to, particular places too strong to disregard on the basis of an unending search for ever lower operational costs. Furthermore, there is a range of public sector businesses which contribute significantly to local and regional economies and simply cannot move easily from place to place. Similarly, the mobility of skilled labour has perhaps been overstated, or at least misunderstood. Turok (2004) demonstrates, for example, that the mobility of skilled graduate labour between UK cities depends more on the employment on offer than on the particular social and cultural amenities or attractiveness of a place. Talk of a flat, borderless world where geography no longer matters is thus overtly simplistic of the variegated reality of local economies that exists, each with its variously embedded forms of economic activity and resources.

A second problematic axiom at the heart of the competitiveness discourse is its understanding that the critical drivers of competitiveness are over-whelmingly and almost exclusively supply-side in orientation, such that little or no attention is paid to the creation of demand for the region's goods and services. Yet a low level of local demand has the potential to dampen local innovativeness and entrepreneurialism and may hinder the development of high-quality cultural and infrastructural capital. Whilst the supply-side may be critically important to business productivity and growth, supply-side measures will fail if insufficient attention is paid to creating or sustaining the level of demand, perhaps through favourable macroeconomic conditions and policies (Kitson et al., 2004).

The regional competitiveness discourse is also characterized by the belief that regional economic performance and prosperity is ultimately derived from and thus is reducible to the competitiveness of firms in the region. The com-petitiveness discourse has asserted that the characteristics of firms can be unproblematically imposed onto regions – from competitive firms to compe-titive regions – which 'manifests a dangerous shift from the rationality of the firm as instrumental actor to the rationality of the region as instrumental actor' (Hadjimichalis, 2006: p. 698). In so doing, it displays a narrow focus on the firm and thus on *growth in* a region rather than the *development of* a region (see Markusen, 1994; Perrons, 2004). Thus, even if the rationale for creating a supportive microeconomic business environment for the develop-ment of productive firms is accepted, the importance of this for the improve-ment of regional living standards is much less clearly defensible. Indeed, Porter's thesis has been criticized for its general failure to perfectly reconcile the micro-level analysis of the competitive advantage of firms with the macro-level analysis of development prosperity (Grant, 1991). As Martin and Sunley (2003: p. 15) observe, 'equating competitiveness with productivity is to invite tautology and ontological confusion: is a region more competitive because it is more productive, or is it more productive because it is more competitive?'

There are a number of difficulties in asserting that simply having a stock of more productive firms necessarily makes a region more prosperous. Firstly, the direction of causation between productivity and regional prosperity is itself problematic. Higher standards of living in a region may attract investment from more productive firms over the longer run, or change the structure of economic activity. Thus, income growth may propel productivity improvement, as well as the other way around.

Secondly, the links between competitiveness and regional prosperity may also be highly contingent upon the character and stock of firms and industries in a region, a fact largely ignored in the competitiveness discourse. Regional productivity is clearly an average of a region's different economic activities, and will therefore reflect its particular industrial structure and pattern of specialization (Markusen, 1994). High productivity firms and sectors may well sit 'cheek by jowl' with low productivity activities. Reinart (1995) also argues that high relative or absolute productivity does not necessarily lead to a higher standard of living – being the most efficient in the 'wrong' (i.e. low value-added) activities, for example, leads to low standards of living. This suggests that it is a region's ability to enjoy rapid *changes* in the level of relative productivity that is more significant in terms of regional prosperity than absolute or relative regional productivity levels per se (Camagni, 2002). Furthermore, economic activity and employment rates also play an important role in determining a region's overall living standards, and these are likely to be influenced for a whole host of economic, social, cultural and political factors and not simply by factors pertaining to firm competitiveness.

A further, related problem arising from this narrow focus on the firm *in* the region is the tendency to ignore the constitutive outside influences, social relations and networks that shape regions and their development processes. The relational perspective on regions (discussed in the Introduction) conceives of regions as open, discontinuous 'spaces of flows', constituted by a variety of social relationships. This perspective sees patterns of regional development and prosperity as reflecting relations of powers and control over space, where core regions tend to occupy dominant positions and peripheral regions play marginal roles within wider structures of accumulation and regulation. In this regard, each regional economy is in a distinct position, since each is a unique mix of relations over which there is some power and control and of other relations within which the place may be in a position of subordination.

Yet, with competitiveness, regional economies are typically conceived as separate from each other, with no serious account taken of the relations between them. This perspective posits that certain regions, such as the South East of England, are likely to develop a hegemonic political and economic position which not only shapes their development but also impacts on the development processes of other regions. As Massey (2007a: p. 106) observes, 'it is the Newtonian, billiard-ball world – here of isolated regions instead of isolated individuals; but it does not reflect the real economy. And it makes it very difficult to entertain the possibility that the growth in one region might have deleterious effects on the prospects of others.'

Regions are often locked into complex interdependencies and networks of relations, sometimes cooperative rather than necessarily conflictual or competitive. They create markets for one another, people sometimes commute between regions, and supply chains often cross regional boundaries. Even the most 'competitive' region achieves that status by winning employment and activities from other locations, and therefore by exporting goods and services to those other locations. Thus, if one region does poorly and cannot supply good-quality or cheap goods and services, in all probability this will not be good news for neighbouring regions, which are likely to suffer the consequences of this reduced supply, as well as perhaps falling demand for their own goods and services. Put another way, there is a dialectic relationship here, whereby the successful region needs, and depends, on the markets of other regions. Hence, we find that all regions in the UK are concerned if, say, London and the South East enter a period of low economic growth, because this will have negative effects on all other regions. Competitiveness thus brings a form of vulnerability even for successful regions and cities, such that the analogy of their acting as business entities, clearly cast as either winners or losers in a dog-eat-dog competitiveness game, is fraught with difficulties.

Similarly, the regional competitiveness discourse ignores the role of national and global forces in shaping regions and their development. As Lovering (1999) observes, there are problems in asserting that improving the international competitiveness of firms in a region has a necessary or logical impact on regional economic welfare. The latter depends not only on regional averages in productivity but also on distributional questions, which may or may not be improved by the globally oriented firms in a region raising their market shares. Thus, macroeconomic regulation and the prevailing welfare regime may also profoundly shape regional development pathways, their degree of inclusivity and, ultimately, regional standards of living (Perrons, 2004). Neither are the differences between places and their diverse evolutionary paths fully acknowledged in the dominant discourse. Factors such as luck, history, geography, favouritism, culture and ruthlessness mean that regions and firms are inevitably destined to pursue very different development trajectories. Yet this diversity is not acknowledged.

The limits to competitiveness (3): pernicious outcomes

Taking a broader perspective on regional development processes illuminates the limits of policy approaches predicated upon the dominant regional competitiveness discourse. Adopting the relational perspective would imply that the problems of less prosperous or weaker regions might be explained by their relationships with prosperous, core regions and broader relations of power rather than simply reflect deficiencies in the performance of their firms or the business environment within which they operate. However, the competitiveness discourse eschews consideration of the relations between regions, focusing only on the imperative of building capacity *within* regions. The responsibility for developing competitive firms and prosperous regions is thus placed firmly with institutional actors and communities within regions, who are therefore also seen as culpable when competitive performance is seen to have slipped. Competitiveness league tables are inevitably seductive for regional development agencies, and the media keen to absorb 'quick and dirty' comparative measures of regional economic performance. However, they clearly carry the inherent danger of stigmatizing lagging regions as failing because of their own deficiencies, when the problems may lie in part in broader structures. The policy consequences are also clear. The result is an overarching focus on building institutional structures such as regional development agencies, to the neglect of a more active interregional policy that may aim to both redistribute resources between regions and control growth in the core with equal if not greater impact (Cumbers et al., 2003).

The dominant discourse also leads to an emphasis on a relatively narrow route to regional prosperity, ignoring the potential for growth and development to be achieved through broader, more diverse avenues. In a critique of the spatial development strategy produced for London in 2004 (the London Plan), Massey (2004: p. 16) notes that the analysis in the plan focuses unerringly on the need to support and encourage the continued growth of financial and business services which are regarded as the chief motors of economic growth and job creation. Furthermore, it places huge faith in the importance of attracting inward investment and 'is pervaded with anxiety about *competition* with other places, in particular Frankfurt as an alternative financial centre'. Indeed, the other face of London as a world city – its cultural mixity – is sometimes mobilized in this competition with other places, as it was in the case of the competition to host the 2012 Olympics. The cultural openness of London has been proffered as an attraction to potential Chinese inward investment (Massey, 2007a). Massey (2004: p. 16) concludes that 'this form of self-positioning represents a significant imaginative failure which closes down the possibility of inventing an alternative politics in relation to globalization'.

The regional competitiveness discourse ignores the possibility that regional prosperity might be achieved by, for example, the development of firms serving local and national markets and not just international ones, or by the development of community or social enterprises which meet broader social and environmental as well as economic objectives. Moreover, not all economic activity is undertaken by private firms. Much economic activity takes place within the public sector, the size of which makes it an important contributor to regional economic performance.

Competitiveness-speak privileges an exchange-relations reading of capitalism and reduces regions, with all their institutional, political, cultural and social complexity to little more than 'nodal points in a spatial cost surface that extends on a global scale' (Ward and Jonas, 2004: p. 2124). As Petrella (2000: p. 70) writes, the discourse instead sees the region as 'a territorial fragment of a global network of flows, resources, services and markets'. It is as if the region has been transformed from a community into an economic ensemble of commodities (goods, resources, services, financial capital and so on). Hence, in turn, he argues, 'regional policy has been transformed into a regional resources management policy where people don't matter' (p. 70). Policies instead tend to prioritize rather narrow, private-sector orientated agendas at the expense of broader regeneration initiatives – a criticism frequently levelled at the English Regional Development Agencies, for example.

In particular, competitiveness strategies do not demonstrate any concern with who benefits from these productive firms and supposedly competitive places, nor indeed from the sustainability of these outcomes. This reflects the narrow way in which the 'environment' for successful firms is defined by this discourse. By focusing on the narrowly microeconomic and ignoring the social and ecological environment, the discourse fails to promote policies encouraging the development of firms which are sustainable and fitted to the broader social needs and environmental limits of a place. The discourse clearly fails to address the question of sustainability or the possibility that the outcomes of relying on a strategy that is based on internationally competitive firms may not necessarily be desirable or help a place to meet its other objectives. Regions do not have a single, purely economistic objective and they are about much more than the pursuit of 'profit' or ever-improving economic outcomes. They may also have other environmental or social objectives, such as the desire to secure a more equitable distribution of income, to provide some minimum level of environmental quality, or improved standards of public services, particularly in relation to health, education and social care – the very real and tangible policy ends that impact on the living standards of their populations.

If these objectives are not met, then over the longer term the situation will almost certainly not be sustainable. If the aim is to increase average earnings in the long run, for example, it is only logical that improving competitiveness should involve alleviating poverty – persistent poverty will almost certainly hold back competitiveness. Furthermore, a locality or region may clearly be competitive today by depleting and denuding its physical environment, but then it will clearly not be competitive tomorrow as a result. But the competitiveness discourse ignores these issues. In promoting the interests of productive firms above all else, it is overtly a growth-first agenda, and yet there is indeed growing recognition that economic growth does not equate to progress in respect of human development and well-being.

This apparent paradox is neatly summed up by Richard Layard (2005: p. 3): 'we are now much, much richer than previous generations. We have more food, clothes, cars and holidays, bigger homes, nicer jobs and better health. Yet even as incomes have more than doubled, we in the West are no happier than we were fifty years ago.' There is indeed a substantial body of literature which now points clearly to the fact that rising levels of GDP in many advanced economies have not been translated into rising levels of subjective well-being or happiness. For example, Jackson (2004) observes that levels of life satisfaction have not risen in the UK for over 30 years.

A number of explanations have been put forward to try to explain this paradox (see Donovan and Halpern, 2002; and Layard, 2005). Central to many of these explanations is the idea that there are factors other than income associated with income growth that offset its positive impact on life satisfaction. For example, greater economic growth may be accompanied by increased 'economic bads' or social costs such as increased pollution, wasteful journeys, growing inequality, rising rates of crime and greater incidence of family breakdown, divorce and social disharmony. Clearly some measures of success hide other kinds of failure. To create more and more growth we need more and more consumption, but this does not lead to happy families and fulfilling lives. Indeed, there are clearly limits to beneficial consumption. This is perhaps best illustrated by the growing obesity problem afflicting western societies. More growth does not necessarily mean more benefits. In fact it creates 'hidden' human, economic and environmental costs which may impede economic prosperity in the long term. Furthermore, individual and collective societal well-being do not come only from material wealth or traded goods. Well-being is clearly also shaped by a number of *non-economic* factors – such as our genes, physical activity, social activities and the strengths of our social and family ties and relationships.

Yet competitiveness is not concerned with the well-being of individuals or communities as a whole. Instead 'the well-being of workers and local residents has been replaced by a discourse that talks exclusively of the well-being of firms and regions' (Hadjimichalis, 2006: p. 700). It rarely takes account of the non-economic variables essential to the social reproduction of everyday life – that is, the ways and means by which people exist and make a living – such as unpaid care giving and volunteering activities and freely available environmental resources, and may in fact wreak considerable damage on them and impose huge human and social costs (Turok, 2004; Jarvis, 2007). Bourdieu (1998) refers to the utopia of endless exploitation propagated by neoliberal competition where all collective structures which may impede the pure market logic are destroyed, with little understanding that such structures and an associated sense of collective responsibility and shared endeavour are fundamental to the reproduction of capital. Competitiveness is ostensibly economically framed and measures quality of life in only narrowly circumscribed economic measures, and pays scant regard to the broader social, ethical, moral and ecological determinants of well-being. As McDowell (2004: p. 146) observes, 'what [competitiveness] cannot do is allocate those resources that are outside the market – goods and services and labour exchanged voluntarily or for love, in the household and in the locality'.

As a final indictment, competitiveness also results in invidious and damaging place-based competition. As has been demonstrated, the policy discourse around competitiveness tends to conflate the strategic pursuit of competitiveness (as meaning the search for improved economic performance) with engagement in competition for resources and the development of boosterist strategies aimed at attracting high-quality, innovative, knowledge-based firms and skilled labour. Competitiveness is a catch-all for the pursuit of business-led growth and entrepreneurial place selling – in short, place promotion. In this regard, the pursuit of competitiveness ostensibly promotes and encourages competition between places around their attractiveness and image. As a consequence, regions (and also cities) are increasingly engaged in a familiar hotch-potch of property, retail and events-led interventions targeted at improving the quality of the business, cultural and social environment. This has a number of undesirable consequences.

The emphasis on competitiveness as attractiveness means, in effect, that regions (and also cities) are increasingly engaged in two somewhat contradictory contests. On the one hand, they are preoccupied with improving their external image for skilled labour, the 'creative classes', businesses and investors. On the other, they are simultaneously engaged in persuading central or federal governments and funding authorities such as the EU of the urgency of the claims for funding assistance to address problems of urban and rural decay, structural change, high unemployment, poverty and social deprivation. Regions thus face an ongoing problem of consistency between 'participating in a beauty contest for people and private investment, and also an inverse beauty contest (an ugliness contest?) for public funds' (Collins, 2007: p. 76).

Regions are, then, perhaps left to proclaim 'success' from securing large-scale programmatic funding for regeneration and renewal which in truth reflects their relative lack of prosperity and economic strength. Alternatively, the dwindling pool of mobile firms often becomes the source of the 'trophies' paraded by successful regions and localities, with little attention paid to the lack of embeddedness of those firms, or indeed the potentially negative consequences of their privileged status. This creates huge vulnerability for regions that become overly dependent on a small number of large multinational firms. This gives large corporations huge power in relation to public policy, with capital able, in effect, to shape significantly the policy choices of governments, disempowering people's ability to set the agenda for their region (Purcell, 2009). It also means that regional strategies are necessarily geared to external audiences, business investors, events organizers and sponsors and so on, and not to resident communities. It also propels growth in competitiveness councils, think-tanks and various other undemocratic institutions that can provide the appropriate toolkits, best practice guides and metrics deemed essential to secure the desired 'success'.

A further problem is that the competitiveness game, once set in motion, perpetuates a seemingly unending policy task – a classic 'Red Queen' effect or never-ending, evolutionary arms race between competing entities.[1] Regions caught up in the competitiveness arms race each end up running faster and faster to effectively stand still, since every other region is engaged in similar activities at an equally frenetic pace. The results can be incredibly pernicious, not least because the price of the key investments and resources being sought simply escalates, thereby cancelling out the benefits of any other innovations or efficiencies.

Furthermore, rivalry between places can lead to wasteful duplication of public services or lost opportunities through an unwillingness to collaborate on more broadly based joint ventures. Swyngedouw (1992) observes that the 'frenzied' and 'unbridled' competition for cultural capital results in over-accumulation. This risks creating the zero or even negative-sum game of entrepreneurial governance (Harvey, 1989), where 'places face the possibility of being caught in a vicious cycle of having to provide larger subsidies to finance projects that deliver even fewer public benefits' (Leitner and Garner, 1993: p. 72). They become impelled to engage in salesmanship and to develop prestige projects so as to try to keep up with their rivals, activities which focus on creating an image that will help to attract in the capital and labour that are deemed so critical to competitive success.

However, such projects do not necessarily bequeath a competitive advantage and may instead work simply to fuel speculative development which is of little import in terms of assuaging economic problems. In short, the pursuit of competitiveness propels an overtly narrow focus on the *promotion* of a region's assets rather than on their *development* (Unwin, 2006). Accordingly, many projects have been shown to be financially unviable in their own terms and have been propped up on public money on the (unproven) understanding that they will lure in global investors. And clearly, in doing so, these projects can take resources and investment away from other budgets such as for housing, education, social care and so on. In this respect the obsession with competitiveness and the desire to out-perform rivals may fuel the uneven development at the heart of capitalism that devalues one place in favour of another (Swyngedouw, 1992).

This potential for uneven development is exacerbated by the fact that the competitiveness race between regions is an inherently unequal one from the start. By way of illustration, the competitiveness agenda that dominates UK regional policy has been sharply criticized by those seeking to engineer development in the North East of England, which continues to suffer a legacy of industrial decay and decline. As a representative from the regional development agency in the region has put it, 'a core element of [government] thinking is that every region should maximize its strengths and address its weaknesses. It's like telling everyone in a race to run faster and expecting people at the back to catch up with people at the front. What [the government] is refusing to do is get the people at the front to run slower so that the people at the back can catch up' (cited in North et al., 2007: p. 16). In the absence of a redistributive regional policy to counter the competitiveness ethos, it is difficult to see how regions such as the North East can realistically catch up. This is especially the case when the economic growth and power of the South East is being reinforced by developments such as the 2012 London Olympics.

The competitiveness discourse instead conceives competition as occurring between places that begin competing on a level playing field, with fortune favouring the entrepreneurial. Yet regions are manifestly not like firms and are instead much broader, often territorially defined, social aggregates, with very different economic and political structures. They do not compete on a level playing field. In fact, each region is embedded in a set of national and regional regulatory systems, institutions and norms, and occupies a unique development trajectory as a consequence of its historical role and location within the broader, evolving political and economic system. This creates differences in industrial structures, levels of maturity and ability to respond to the forces of economic and political restructuring. Moreover, higher levels of the state frequently exercise political favouritism, either deliberately through spatially targeted policies or as the unintentional result of national policies with different regional and local impacts (Sheppard, 2000). These dimensions of difference all inevitably serve to tilt the playing field to favour some places over others, and to ensure that certain regions have an innate advantage in the competitiveness game.

Conclusion

The discourse of regional competitiveness is thus hugely problematic and limiting for regions in terms of policy and practice. It is a narrowly reductionist and deterministic discourse which forms a particular view on the route to regional development, one which is based purely around the economic growth of particular firms, serving particular markets and in particular, bounded spatial configurations. In line with its embeddedness within a globally neoliberalist rationality, it focuses on the narrowly quantifiable economic aspects of regional economic performance. It thus ignores the inherent diversity of local and regional economies and seriously understates the significance of building local well-being on the basis of assets that extend beyond the narrowly microeconomic. Moreover, its strategies appear to have failed to reduce inter-regional inequalities and may in fact be helping to widen them, inasmuch as the discourse of competitiveness has encouraged an overt policy emphasis on the institutional and business conditions for ensuring growth *within* regions at the expense of a broader concern with the interdependencies and relationships *between* regions. It is thus both highly tendentious and flawed, and not only betrays a serious failure to understand how regional economies actually work, but also, as a final indictment, results in invidious and damaging place-based competition. In short, competitiveness, it seems, is at one and the same time banal and commonsensical, common-place and everywhere, as well as unsustainable and 'care-less'.

This is not to dismiss the value of competition outright. It can clearly be a positive experience to learn from others and to seek to improve overall standards. But it is narrow, place-based competition, concerned with the place promotion of the narrow pursuit of particular kinds of firms and people with a growth-*only* intent that is the problem. The discourse of regional competitiveness necessarily paints a picture whereby the dynamics of competition are shrouded overwhelmingly in positive rather than negative connotations. Neoliberalism seeks to establish a particular common-sense notion that competitiveness is an intrinsically good thing, an ethic that will help to generate wealth and ensure happiness. To the extent that this ideological project has been successful, it has narrowed the policy options available to regions, to governments generally, and to the people they represent. As Purcell (2009: p. 145) observes, 'a polity that values the environment, for example, might feel it cannot make a strong environmental policy (e.g. signing on to Kyoto) because it would make the area less competitive. The neoliberal claim is that competition is a question of life and death.' Regions feel that they must be competitive or die. Other strategies look very optional in the face of the competitive and global struggle for survival.

However, when the larger spaces within which spatial competition occurs, and the uneven development that typifies economic differences between regions, are introduced into the analysis, the dynamics of competition appear more negative than positive (Sheppard, 2000). As Backlund and Sandberg (2002: p. 90) observe in relation to a study of new media firms and networks, 'research has been suffering from a success bias, primarily concerned with explaining why the winners win and not why the losers lose'. More empirical research needs to be directed towards identifying what options there are for regions that do not have the cultural and institutional conditions conducive to the development of innovative, internationally successful firms. In the absence of this more rounded view of the different modalities of regional competition, policy will continue to be based on the rather naive assumption that everyone can be a winner.

Furthermore, in asserting that regions are bounded, discrete, separate spatial entities, the discourse ignores the fact that regions are locked into increasingly complex interrelationships and interdependencies which create particular vulnerabilities for them, especially where they are over-specialized in particular goods and services. These vulnerabilities are becoming ever more exposed in a world characterized by increasing uncertainty over the availability and security of resources, particularly in relation to food, oil and energy more broadly. This implies the need to better understand regions as spaces of flows and to acknowledge the more complex political economy within which regions function. This draws attention to the complex interplay between the bounded, territorial nature of regions in administrative and governance terms, with the more complex relational spaces of regions in broader economic and ecological development terms. This suggests the need for a new paradigm for understanding regions – a paradigm which moves us beyond competitiveness. It is to this issue that the final chapter now turns.

7   Resilient regions

Re-'place'ing regional competitiveness

Introduction: the seeming absence of any alternative

The competitiveness discourse clearly has a demonstrably pervasive grip on the thinking, policy and practice around regional development. It infuses much of the theoretical debates that have been struggling to catch up with and rationalize policy practice; it shapes the metrics and indicators of regional performance and 'success' which are used to help spread and sustain its logic; it dominates the strategic imperatives for regions; and acts to subjugate and constrain the choice of policy approaches pursued. Quite simply, competitiveness has been elevated to the status of a natural law in the modern capitalist economy.

The appeal of place competitiveness to policy makers at all spatial scales is easily understood when considering competitiveness ostensibly as an idea, promoted by particular and powerful interests. The very vague and nebulous nature of the competitiveness concept can then serve to act as one of its principal strengths for policy makers. It becomes a veritable garbage can into which all relevant strategic actions for the support of particular goals can be thrown. For cities and regions, the most striking characteristic of competitiveness strategies is indeed their extreme vagueness and flexibility over time. In the event, the commitment to competitiveness can thus be interpreted as authorizing key players in urban and regional governance to adapt to circumstances and opportunities as and when they arise, and especially as they have been proposed by potential investors, notably in the property development sector (Bristow and Lovering, 2006). At times, competitiveness can also be used strategically to justify the support of particular ailing local businesses, perhaps for political purposes or electoral gain. At other times, the cloak of competitiveness usefully absolves local policy makers from responsibility, inasmuch as the winds of global competition and change can be conveniently presented as variables which are essentially outside local responsibility and control.

Perhaps more disturbingly, however, the notion of competitiveness as an idea backed by powerful interests also highlights its potential to achieve hegemonic status such that, in spite of its limitations, it becomes very difficult to conceive of any alternative, at least at the level of discourse. Competitiveness has become a banal, common-sense notion – competitiveness is simply 'the way it is'. Indeed, as Fougner (2006: p. 169) observes, 'the problem of international competitiveness [is] a seemingly given one that states cannot but attend to. There is seemingly no alternative but to compete – unless, that is, a state should be so unfortunate as to be excluded from the competitive game altogether for structural reasons – and thus little in the literature pointing beyond the competition state.' Whilst alternative discourses around cooperation have been conceived (see, for example, Bunzl, 2001) and progress made in the development of alternative economic spaces, these have as yet failed to induce any systematic or global transformation in the dominant thinking.

The obdurate hold of competitiveness thinking is not surprising, given its central place in the dominant force that is neoliberalism. The ideology of international competitiveness is propagated globally and is predicated upon the contemporary dominance of a neoliberal rationality of government. This is characterized by the constitution of the market as the ideal in relation to which governance should be orientated, with governance rationality derived from arranged forms of entrepreneurial and competitive behaviour, legitimized where there is perceived to be market failure or imperfection (Fougner, 2006). As George (1999) states, 'the central value of ... neoliberalism itself is the notion of competition – competition between nations, regions, firms and of course between individuals. Competition is central because it separates the sheep from the goats, the men from the boys [*sic*] and the fit from the unfit. It is supposed to allocate all resources, whether physical, natural, human or financial, with the greatest possible efficiency' (cited in McDowell, 2004; p, 146). Thus, the promotion of competitiveness is critical to the maintenance and reproduction of the capitalist hegemony within and beyond advanced capitalist countries (Cammack, 2006).

Within this context it evidently becomes very difficult for individual nations to opt out of the competitiveness game without being seen as attempting to manage expectations downwards or to lack ambition. As Layard (2005) points out, the first worker to suggest shorter trading hours for lower wages is likely to have his commitment questioned. The same can be said for individual governments at national, regional and city scales, particularly given that perceptions of economic management remain one of the key overriding drivers of electoral success. Competitiveness-speak may create a trap for policy makers from which, once entrapped, they struggle to break free. To fail to pursue competitiveness or to opt out of the competitiveness game completely, might be regarded as either a sign of weakness or a submission to defeat by erstwhile rivals.

It would thus appear as if there is no alternative for places but to adopt competitiveness-orientated policies. The story of inescapable globalization, competitiveness and economic neoliberalism is indeed a very pervasive one, such that even ideas that intuitively appear to challenge its orthodoxy are subsumed to its clutches. One might argue, for instance, that the growing interest in the role of cultural creativity in cities has been hijacked by the competitiveness project inasmuch as the dominant thinking (propelled by key authors such as Richard Florida) sees them as ostensibly being about developing greater place attractiveness (competitiveness) and not about celebrating the uniqueness and diversity of local and regional economies per se. Similarly, there is evidence that welfare and social policies are being reconfigured to support economic competitiveness and the pursuit of growth (see Chapter 5). As Massey (2000: p. 282) observes, 'every attempt at radical otherness [becomes] quickly commercialized and sold or used to sell ... With all of this, one might as well ask what are, and where are, the possibilities for doing things differently?'

The growing awareness of the limitations of competitiveness-oriented policies for regions makes these questions especially pertinent. Furthermore, the CPE approach used throughout this volume provides a useful framework within which to begin to answer them, since, as has been demonstrated, it provides revealing insights into the ways and means by which prominent ideas are spread, the political dynamics of their evolution, and their utility for advancing particular policy goals and discouraging others. In emphasizing that competitiveness is a social and political construct, the CPE approach also usefully highlights that it is not immune to challenge and resistance. Whilst economic imaginaries such as competitiveness can be discursively constructed and materially reproduced at different sites, and on different scales, they are only ever likely to be partially constituted and will remain contingent. As Jessop (2005: p. 146) observes, the process of material reproduction 'always occurs in and through struggles conducted by specific agents, typically involves the asymmetrical manipulation of power and knowledge, and is liable to contestation and resistance'. This means that there will always remain space for competing imaginaries to challenge the dominant ones and, as this volume has demonstrated, actual practices may vary significantly in the domains of policies and institutions. In other words, competitiveness is recontextualized as it is rolled out, replicated and reproduced across different spatial scales. Context and scale thus appear to matter, in part at least.

The possibility that competitiveness may be contested raises a number of critical sets of questions about the sort of alternative imaginaries that might emerge, how and whether they constitute an antagonistic challenge or countermovement to competitiveness, and from where they are likely to emanate. First, given its foundational role in cracking the whip of competitiveness and setting the extra-local rules of the game, there are fundamental questions to be asked regarding the apparent immutability of neoliberalism and the scope which exists to effect a fundamental transformative shift in its status, its privileging of economic rationality and the imperative of competitiveness at the grand level or global and national scale. This raises questions regarding the catalytic forces for change – forces which appear to be gathering momentum in a global economy increasingly characterized by economic instability and tangible concerns regarding the material limits of capitalism.

The second set of questions relates to the alternative discourses or economic imaginaries which might be conceived for regions. The questions here concern what form these imaginaries might take; the extent to which they are radical counterpoints to, or simply weakly modified variants of, dominant thinking; and from where they emerge. There are particular questions around the precise role of the 'region' in the development and spread of these imaginaries. The region has been a determinate 'space of competitiveness' and a key transmission mechanism for the reproduction and recontextualization of the competitiveness discourse. Questions thus surround whether it fulfils a similar role for the shaping and spread of alternative imaginaries. Do these imaginaries develop and prosper within the agonistic, political struggles and practical, policy-focused debates which are showing signs of emerging within regions? And to what extent do they resonate with and draw insights from the nascent, theoretical rethinking of how regions and regional development processes actually work?

The third and final set of questions concerns the permissive, regional and local factors shaping whichever (if any) of these alternative imaginaries might grow to command the more widespread degree of rationality required to propel interest-related action and to effectively challenge the competitiveness hegemony within the regional development discourse – in essence, how might a transformation 'from below' be effected. This focuses attention on where the power lies to effectively challenge dominant imaginaries and what mechanisms might be deployed by relevant actors to effect their wider co-option and spread.

The purpose of this chapter is thus to consider each of these questions in turn. It begins by examining the edifice that is neoliberalism.

Contesting neoliberalism

There are good grounds for asserting that the apparent immutability of neoliberalism has been somewhat overstated. It is without doubt an epochal political and ideological project that has come to describe a fundamental and far-reaching shift in political economy from welfare-based government to competitiveness-driven governance. It is a global project aimed at establishing and spreading the dominance of certain key assumptions, namely the supremacy of market forces and free trade, the undesirability of widespread state intervention, and the necessity of labour market flexibility. As such, it is itself a discourse, a social construct, and one which is structured essentially by multinational corporations (Peet, 2001).

As with any hegemonic regime, its concrete and ideological elements require continuous refortification. In other words, cracks and instabilities emerge as a matter of course because it produces its own problems of contradiction and legitimacy, and it must always articulate with and, to an extent, accommodate existing policies, habits and assumptions (Purcell, 2009). Indeed, various authors have drawn attention to the varieties of neoliberalism that emerge as it is rolled out across national political economies with their differing contexts, conditions and challenges. For example, Birch and

Mykhnenko (2009) demonstrate how deregulation, privatization and trade liberalization processes have been pursued for different reasons, in different ways and to varying extents in the economies of Western and Eastern Europe. This has led to the variegated restructuring of regional economies – a process which is in itself revealing of neoliberalization as a process, and not simply as a discourse. This clearly echoes one of the principal themes of this volume, namely how the process of rolling out or implementing competitiveness at the regional level is fitful and uneven and how competitiveness becomes recontextualized and is rendered variegated and hybrid as it comes into contact with the specificities of particular regions and places.

Neoliberalism has indeed demonstrated an obdurate and enduring capacity to reinvent itself by accommodating criticisms and co-opting them within its logic. This is particularly pertinent in relation to the emergence of green capitalism, which has been criticized as 'nothing less than a major strategy for ecological commodification, marketization and financialization which radically intensifies and deepens the penetration of nature by capital' (Smith, 2007: p. 17). Its various manifestations, such as carbon credits and eco-industrial parks are thus often regarded as perpetuating the same neoliberal logic and route to development which is merely refreshed with a slightly green tinge (Bristow and Wells, 2005). It thus remains an open question whether sustainable capitalism is possible. As Hudson (2005) argues, whilst there are strong reasons for believing that some form of eco-capitalism is improbable, the chances of a systemic non-capitalist alternative appear at least equally distant. This leads him to conclude that 'the impossible may indeed be necessary and the necessary impossible in seeking to create sustainable economic practices, flows and spaces – at least in the absence of some radical re-appraisal of concepts such as "the economy", "productivity" and "development" and consequent changes in practices linked to a fundamental change in the social relations that dominate the economy and define legitimate practices' (p. 250). What is needed is nothing less than a 'total' solution to the contradictions and crises of capitalism, through a set of social relations that produce a more equitable and sustainable economic, political and environmental settlement (Harvey, 2000).

Whilst the probability that neoliberalism itself will be toppled may therefore be unlikely, the scope for transformation and change should not be entirely ruled out. Indeed, recognizing the multiple and often contradictory nature of neoliberal spaces, technologies and subjects in practice points to the potential scope which exists for resistance and challenge. Moreover, the potential for major catalytic forces to fundamentally destabilize neoliberalism may be gathering momentum in the form of the immediate global financial crisis which is spawning a new age of austerity; climate change which is leading to extremes of, and greater instabilities in, weather conditions and provoking significant change in the relative viability of different economic activities in different places; and the end of the era of cheap and plentiful oil, with all its attendant implications for the longevity of carbon-fuelled economies, cheap, long-distance transport and global trade in material goods.

This 'triple crunch' powerfully illuminates the harsh and potentially disastrous material consequences of the voracious growth imperative at the heart of neoliberalism, both in the form of resource constraints and in the inability of the current global system to manage the financial and ecological sustainability of the global economy. In so doing, it has the potential both to galvanize previously disparate, fractured debates and protests about the merits of the current system and to encourage the search for alternative, more coherent visions for the economy and society, and for governance and policy. As such, it clearly has the potential to change public and political opinion in favour of a new, global concern with frugality, egalitarianism and localism (see, for example, New Economics Foundation, 2008; Jackson, 2009).

Precisely how the obstacles of capitalist and state power are to be surmounted and dismantled remains unclear and will necessarily form the key question to be addressed by advocates of change. Yet it is the very opening up of a debate and the countenance of the possibility of alternatives, of change, that is nonetheless a hugely important first step. As Purcell (2009: p. 155) argues, 'resisting neoliberalism requires movements that can actually deliver a fully expressed "radical and transformative" politics'. What is also clear is that these pressures highlight that it is wrong to talk about neoliberalism as some unstoppable force which is automatically and unrelentingly transmitted from place to place. It is a construct and, as such, 'what was made can be unmade' (Leitner et al., 2007: p. 325). There are indeed agents who resist or struggle mightily against its power, such that whilst it is hegemonic, it is neither invincible nor everlasting (Massey, 2007b; Leitner et al., 2007). Instead, 'it is merely hegemonic *now*' (Purcell, 2009: p. 144).

Alternative economic imaginaries and spaces

Recognizing the vulnerabilities, variations and contradictions inherent in neoliberalism also draws attention to the potential scope that exists for constructing 'alternative' economic imaginaries which are grounded in alternatives to the social relationships of mainstream neoliberalism and its competitiveness imperatives. In the contemporary neoliberal restructuring processes propelled by hegemonic capitalist forces globally, the rolling out of new policies and regulations on the ground inevitably results in their collision with existing, specific structures of governance and procedures. Institutional change is thus path dependent. This tends to create institutional compromises and hybrid policy frameworks, such that whilst neoliberalism remains *discursively* hegemonic, *actual policy practices* in the domains of policies, institutions and norms remain a much more variegated patchwork of institutions, agencies, policy agendas and governance relations. As such, they inevitably contain and include contradictory social forces and challenging political movements (Moulaert et al., 2007).

Within a fragile and susceptible capitalism, a wide range of 'alternative' ways of imagining and performing economic activities has indeed proliferated which may provide scope for transformative development 'from below'. Contestations of neoliberalism draw on a range of what might be labelled 'non-neoliberal' imaginaries, which are simultaneously social and spatial and which imagine alternative visions of justice, democracy and ecology – of living and working 'differently'. Many of these emerge at the sub-national, especially sub-regional or local scale, and include alternative systems of finance and exchange such as Local Exchange and Trading Systems, time banks and credit unions; alternative business forms and practices such as social, not-for-profit enterprises, cooperative ventures and short, localized forms of food supply chain such as organic box schemes and farmers' markets; and various projects that seek to enhance environmental quality and revalorize the built and natural environments, or induce more ethical and sustainable practices of consumption (Gibson-Graham, 1996; Leyshon et al., 2003; Hudson, 2005).

These 'alternatives' are inherently diverse but share some commonality in being shaped and directed through sets of social relations differentiated from – and in some cases opposed to – mainstream relations, and thus centred much less around property rights and rights to capital accumulation and unlimited growth. They are also anchored in an emphasis on the rights to the satisfaction of human needs, to cultural emancipation and political and social empowerment, and to the maximization of use value rather than exchange value (Moulaert et al., 2007; Purcell, 2009) – ostensibly, then, to development rather than growth. These are in turn inherently connected to a range of emerging practices and processes of contestation and opposition to capitalism which flourish around the world in the form of anti-globalization protests, organized collective and social action and various forms of militant particularism (Harvey, 2000; Leitner et al., 2007).

Requisite variety or radical counter-movements?

Critical questions surround the potential for these diverse localized struggles to effect more widespread, transformative or even counter-hegemonic change (Leyshon et al., 2003; Leitner et al., 2007). There are good grounds for questioning the extent to which many of these experiments represent a meaningful departure from neoliberalism, with several authors concluding that these are ostensibly progressive moments in the development process which ultimately represent a further embedding of privatized power and the de-politicization of potential conflict. For example, in a review of various 'change movements' across Europe which are challenging the status quo or the move towards further commodification of public use values, Moulaert et al. (2007: p. 207) conclude that 'most initiatives [are] not very powerful in developing a counter-hegemonic discourse', principally because they are 'missing the heavy-weight political and financial authority which could have empowered them to receive sufficient visibility and become hegemonic'. They typically remain 'local experiments on the fringes, or in the interstices of the mainstream capitalist economy rather than systemic alternatives to it' (Hudson, 2005: p. 250).

Furthermore, as Leyshon et al. (2003) demonstrate, the relationship between these localized practices and the 'other', namely mainstream forms of capitalism remains unavoidable, ambivalent, unequal and full of contradictions. Some alternatives, such as LETS systems, appear more radical than others and propose nothing less than the displacement of established norms and practices through the creation of separate economies with separate means of valuation, practice and regulation. Others, such as credit unions and various forms of social market, may be regarded as essentially there to 'mop up' or 'prop up' the mainstream. All alternatives also run the risk of being incorporated into the mainstream and supporting it by providing it with various material and social supplements which help to legitimize it. In other words, they provide evidence that mainstream capitalism is tolerant of diversity and difference, without mounting a serious challenge to its material or social power (Leyshon et al., 2003). Indeed, as Chapter 5 has demonstrated, the discursive power of competitiveness is such that it is capable of framing and co-opting many kinds of counter-hegemonic response, thus creating a sort of 'requisite variation' which reinforces the durability and spread of the overarching imperatives.

However, the power of these alternatives lies as much in the possibility of their existence and their demonstration that there are alternative forms of practice in evidence (Leyshon et al., 2003). Mobilization of such practices at the sub-national scale has thus enriched the repertoire of progressive alternatives to the market and to consumer monoculture (Leitner et al., 2007). But it has also illuminated the constraints and difficulties which may inhibit the realization of their full potential, notably their tendency to exhibit divergent strategic and political priorities, and their often limited political capacities and resources. This suggests that critical factors in the transformative potential of alternative practices will be their capacities to come together around a unifying imaginary and their ability to command greater political authority and resources. Focusing on what unites rather than what divides non-neoliberal practices is potentially very important, inasmuch as it opens up new opportunities to examine the specific ordering and spatiality of projects required to effectively challenge conventional approaches (Holloway et al., 2007).

Resilience

In this regard the imaginary of resilience is of growing interest, as there appears to be a burgeoning coalition of interests and initiatives around it. Resilience has acquired particular resonance and power in respect of the current climate of upheaval and change and the perceived need for places to reduce their vulnerability to the shocks, instabilities and crises of various fuel, food and other resource constraints wrought by the destabilizing effects of economic and environmental turbulence. Various bottom-up initiatives have at their heart the desire to reconstitute local and regional worlds as more resilient spaces based on more socially oriented, smaller-scale and place-based approaches to development. These take various forms such as the development of grassroots democratic and life-centred (instead of profit-centred) economic practices, as in community economy initiatives. Similarly, community agriculture and barter economy movements are focused on creating more robust economic and social spaces by empowering producers and consumers to interact locally, seeking to reduce dependence upon distant and larger-scale agents, namely non-local and large corporations and the nation-state (Leitner et al., 2007).

The building of resilience features most prominently in the Transition Towns movement, which seeks to encourage communities to explore methods for reducing energy use, and of increasing their economic, political and social self-reliance through a variety of place-based initiatives. These range from the development of local currencies to encourage the reduction of food miles and support local businesses through the retention of wealth and spending power locally (e.g. the Totnes pound in Devon, UK), as well as community gardens for the production of food, and various initiatives supporting waste reduction, reuse and recycling. The movement is also explicitly evangelical in its philosophy and places a high priority on national and international networking to disseminate and spread its core ideas. There is some evidence to suggest this is indeed working. The movement has spread quickly from its initial origins in Kinsale, Ireland and there are now estimated to be over 100 communities recognized as official Transition Towns across the UK, Ireland, Canada, Australia, New Zealand, the US, Chile and parts of Europe.[1]

The Transition Towns movement defines resilience as 'the capacity of a system to absorb disturbance and reorganize while undergoing change, so as to still retain essentially the same function, structure and feedbacks' (Hopkins, 2008: p. 54). In essence, it is about 'being more prepared for a leaner future, more self-reliant, and prioritizing the local over the imported' (p. 55), or about closing the economic loops where possible. The meaning of resilience in this context draws heavily on the way in which it is used in the study of ecology. Here it refers to the ability of an ecosystem to withstand external shocks such as forest fires, and to adapt and respond to them rather than simply to wither and die. In this regard, resilient places require a number of key features.

Firstly, they require diversity (as opposed to uniformity) in the number of 'species' of business, institutions and sources of energy, food and means of making a living. Thus, in the context in which supermarkets only have approximately one day's worth of food at any one time, resilience suggests the need to reinforce the local food production culture and diversity of local businesses owned by local people, so that if supplies are stopped from coming in from the outside, the bulk of what is needed can be provided locally.

Secondly, resilient places require modularity, or the capacity to reorganize in the event of a shock, such that they can supply their core needs without substantial reliance on transport and travel, whilst still being linked into each other. In other words, so as to be resilient, places need to be engaged with the wider world, but in an ethic of networking and information sharing, rather than of mutual dependence.

Thirdly, resilient places need to be characterized by an emphasis on small-scale, localized fit-for-purpose activities which are suited to and embedded in the strengths and capacities of the local environment and are aware of and adapted to its limits. This, in place of expensive, large and sometimes predatory or invasive infrastructures, business and bureaucracies. Entire towns or even regions can be left stranded by the demise of key industrial sectors, as has happened with the decline of traditional heavy industries of coal mining, steel and shipbuilding in previous monoculture regions. The experience of Ireland, which built its 'Celtic Tiger' brand and reputation for rapid economic growth in the late 1990s on the basis of inward investment from finance multinationals, but which has been hit particularly hard by the current global recession, presents a further case in point. Thus, as well as diversity across a range of sectors, resilience also implies that these sectors should be relatively small scale, such that no one sector or company becomes locally dominant. In turn, this also implies that resilience is characterized by dispersion (rather than centralization) of control over systems.

Finally, by virtue of requiring mutual use of local assets, capacities and resources, and localized production, trading and exchange, resilience also implies a healthy core or supporting economy of family, neighbourhood, community and civil society, strong in reciprocity, cooperation, sharing and collaboration in the delivery of essential services, care provision and caring of families (Hopkins, 2008; Simms, 2008; Jackson, 2009).

With these characteristics, it is not surprising that the imaginary of resilience is developing a powerful discursive appeal for places on a range of scales. The global 'credit crunch' and the accompanying increase in livelihood insecurity have clearly illuminated the advantages of those local and regional economies that have greater diversity and/or a determination to prioritize and effect more significant structural change (Larkin and Cooper, 2009). Like competitiveness, which has an innate evolutionary survival-of-the-fittest metaphor, resilience has the advantage of drawing on the intuitive and positive connotations of ecological development for spatial economies. The notion of resilience is also sufficiently abstract to be capable of commanding broad appeal across different interest groups, none of which can disagree with its fundamentally positive ethos of improvement and its innate capacities of flexibility, adaptability, recoverability and overall strength. More fundamentally, it appears to address many of the limitations of the competitiveness discourse in its emphasis on more diverse routes to development which are attuned to both the local and the global; its capacity to embrace material and non-material concerns, including the durability and sustainability of development trajectories, and the well-being and social production of people and places; and its capacity to remove the imperative for the constant attraction of external resources and pernicious rounds of place promotion and competition. In short, greater self-reliance or resilience appears to be the exact antithesis of competitiveness (Bristow and Wells, 2005).

Re-'place'ing regions

The growing practical interest in non-neoliberal or alternative economic practices and imaginaries such as resilience raises numerous questions around the spatiality required for them to challenge conventional approaches. In particular, the ostensibly 'localized' nature of alternative practices leads to questions around the role of the 'regional' space in their evolution, progression and development. To what extent are these alternative practices locally or regionally constructed and embedded, or, to put it another way, how do scale and place matter?

Rethinking regional spaces

This focuses attention on the emerging challenge for theorists in understanding a range of questions around precisely the role and function of 'regions'. These include what and who regions are for; more normatively, what they *should* be for; what sort of activities they should encourage; how they are constituted; and how they are connected to other scales and wider geographies. The problem with the competitiveness discourse has clearly been its tendency to reify the region as a space-economy and to assert a narrowly reductionist understanding of its function in ostensibly globally oriented, macroeconomic terms. In so doing, it has failed to see regions as constitutive of multiple, often complex and competing alliances, institutions and interests, and as political, social and environmental as well as economic nodes and spaces. In short, it has perpetuated a rather partial view of the processes of both regional development and state territoriality and strategies of management.

This view is increasingly being challenged and there is an emerging interest in understanding and theorizing the broader and more qualitative character of regional 'development' as opposed to simply economic performance or growth. Pike et al. (2007) provide an insightful and cogent summary of recent debates and highlight the growing diversity that characterizes conceptions of regional and local development. Thus, they assert that 'no singularly agreed, homogeneous understanding of development of or for localities and regions exists. Particular notions of "development" are socially determined by particular groups and/or interests in specific places and time periods. What constitutes "local and regional development" varies both within and between countries and its different articulations change over time' (Pike et al., 2007: p. 1255).

Furthermore, as well as being diverse, regional development trajectories and processes are also increasingly understood to be evolutionary, context specific and path dependent. As Pike et al. (2007: p. 1258) observe, 'from Hackney to Honolulu to Hong Kong, each place has evolving histories, legacies, institutions and other distinctive characteristics that impart place dependencies and shape – *inter alia* – its economic assets and trajectories, social outlooks, environmental concerns, politics and culture'. These factors critically determine their capacity to attract and embed globally competitive activities, as well as their scope for developing alternative formal and informal activities that may be more localized and resilient. These factors also shape the available ecological limits to particular economic activities and development paths, and determine the nature and scale of the resource constraints and challenges of collective consumption and social reproduction that each particular region faces (Weaver and Hudson, 1995; Bristow and Wells, 2005).

Regional development is also characterized by a developing qualitative character which reflects a growing concern with the nature of development, in respect of, *inter alia*, the forms of growth pursued (perhaps 'smart' and 'sustainable'), the quality of jobs created (better equipped to provide 'living' wages and greater opportunity for more socially beneficial work–life balances), the ecological sustainability of outcomes, the embeddedness of investments, and the social and sectoral diversity of new firms (see, for example, Blake and Hanson, 2004).

Qualitative approaches also focus on normative concerns informed by specific principles and values which are socially determined in particular regional contexts and at particular times. These principles and values shape how specific social groups and interests in particular regions define, interpret and articulate what is meant by regional development. They may be collectively held unanimously, shared with a degree of consensus, or, as is more likely, be subject to contest and different interpretations. They reflect the relations and balances of power between state, market and civil society and are socially and politically determined within regions, emerging either from universally held beliefs, held independently of a country's levels of development (such as democracy, equity, fairness and solidarity), or framed by specific perceived economic, social and political problems and injustices. Particular individuals and institutions may also seek to impose their particular interests and visions on others, although these may be contested and challenged. In short, the character, form and nature of regional development is inherently heterogeneous and proliferative, embracing multiple social, political, ecological and cultural concerns as well as the narrowly economic (Markusen, 2006; Pike et al., 2007).

Finally, regions also form defined spaces, if not necessarily strictly territorial or bounded ones, within which particular definitions and kinds of regional development are articulated, determined and pursued. In other words, they remain integral sites and spaces for the formation of rooted understandings, interests and values and the exercise of agency and power. As has already been demonstrated, regions are increasingly conceived as open, porous and ostensibly relational entities, subject to and involved in power relations and thus locked into complex interdependencies with other regions and spaces. This accords with the reality that a particular region's 'success' is often secured by winning employment and activities from others or by exporting goods and services to their markets. It acknowledges the dialectic relationships between regions and the potential vulnerability these bring even for successful locations which, as a consequence, face new geographies of responsibility (Massey, 2004; 2007a). However, in many respects they also remain territorial and bounded entities too.

Thus, 'while flows of ideas, people and resources remain integral to territorial development processes, the expression of localities and regions in which different kinds of development may or may not be taking place in specific time periods is often as territorially bounded units with particular administrative, political, social and cultural forms and identities, albeit those boundaries are continually being reworked and constructed anew at different spatial scales' (Pike et al., 2007: p. 1258). Regions remain important containers for a range of state, quasi-state or non-state institutions and various coalitions of capital, labour and civil society. They also remain important sources for the growth of such institutions, inasmuch as they appear to provide a convenient or appropriate scale for the organization and articulation of a range of economic, political, social and environmental interests (see, for example, Chapter 5).

Where regional governance structures are vested with state powers, new spaces for the agonistic politics required to challenge conventional conceptions and trajectories of development also appear to emerge (see Bristow et al., 2008; and also Chapter 5 of this volume). In allowing more dissonant voices into the policy process, regional governance structures mobilize the sense that policy can be done differently. They also help to strengthen a politics of place and identity, and create a space within which alternative imaginaries and conceptions of development can be articulated and debated.

The capacities of regional governments and governance structures to act on these demands are clearly highly variable and often heavily circumscribed and constrained. Nevertheless, degrees of power exist for regions to make their own political choices in response to these demands (Hudson, 2007), including the proactive search for new metrics of development, and strategic and material support for alternative, non-neoliberal values, objectives and practices. Their unique place in nested structures of multi-level governance also gives them a degree of power (if limited and variable) to act as advocates of change in networks and organizations of national and international government, governance and policy.

Resilience and scale

The spatiality required of the growing number of practical, non-neoliberal or alternative economic practices and imaginaries such as resilience is equally increasingly conceived as rather fluid, multiple and contingent. Alternative social imaginaries typically articulate a variety of geographic imaginations – sometimes seeking to reinforce nationalist imaginaries and the buttressing of the nation-state, but more often than not involving the reinforcement of local economic, cultural and ecological identities and practices as alternatives to a seemingly globalized, borderless, 'place-less' world (Hines, 2000). Moreover, they often explicitly seek to *create* alternative economic, social and political spaces which are in essence more localized, more rooted in and attached to place. Localization thus represents a potentially powerful overarching strategic focus and vision for non-neoliberal practices, but it is not without its limitations. At its most general 'it can buttress a political cosmology in which the very terminology of "localness" carries with it an implication of goodness and warmth' (Massey, 2007a: p. 166). Furthermore, in portraying the local as a product of the global, it carries the implicit danger of territorializing space and creating an overtly narrow and simplistic politics based upon defending the authenticity of the local against the global, when in reality the global itself is produced in local places (Massey, 2007a). The critical point is that implicit conceptualizations of space are crucial in practices of resistance and the building of alternatives (Massey, 2005).

This is affirmed and illuminated by the experience of the Transition Town initiative which, in playing a powerful role in propagating the imaginary of resilience, at the same time provides a revealing insight into changing conceptions as to the role of space more generally and of local and regional space in particular.

The concept of resilience propounded by the Transition movement is rooted in the overarching vision of localization and the search for a progressive, more place-based alternative to development. By definition, it emphasizes and supports the development of activities in the realms of food production, energy use, travel behaviour, retailing and consumption that are more attuned to a place's particular resources, strengths and assets and which reduce the dependence on imported supplies of energy and other resources. Inevitably, then, it is, in part at least, actively seeking to defend the local against the global or, as Hopkins (2008: p. 73) puts it, to encourage 'a return of the local and the small-scale, and the turning away from the globalized'. However, Hopkins falls short of idealizing the local and asserts that 'this will not be an isolationist process of turning our backs on the global community. Rather it will be one of communities and nations meeting each other not from a place of mutual dependency, but of increased resilience' (p. 71). The world is not equally endowed with minerals and raw materials and so it makes sense for places to be able to specialize. The Transition movement thus does not advocate complete localization, but rather partial localization centred on enhanced capacities for local production. The bottom line emphasized by its advocates is that with the onset of peak oil, localization is no longer simply a choice – it becomes an inevitability. In this respect, early preparatory action to enhance local productive capacities, it is argued, clearly makes for good sense.

The Transition movement thus appears less vulnerable to the critique of spatial fetishism, at least inasmuch as it asserts the importance of more complex spatial configurations of activity which go beyond the local, to simultaneously embrace various local–global relationships and connectivities. Resilience is defined carefully, in terms specifically designed to avoid the accusation that its advocates are seeking to construct fences around communities and to create self-sufficiency or a 'nothing in, nothing out' economy. Instead, it is defined in terms of securing a better overall balance or interaction between the 'local' and extra-local and between different nations and areas of the world or, in Hopkins' words, 'more equitable and useful ways of relating in place of the unequal exchanges of "stuff"' (Hopkins 2008: pp. 69–70).

Furthermore, this relational conception of space extends to its own practices and activities as an increasingly global network for mobilizing 'the viral spread' of the transition and resilience concepts (Hopkins, 2008). Transition movements dedicate themselves to sharing their visions, ideas and practices 'so as to more widely build up a collective body of experience' (Hopkins and Lipman, 2008: p. 6), encourage a momentum for change, and work supportively and thus more effectively with other initiatives and groups with similar value systems. In particular, increasing reliance is placed on various forms of organic, internet-enabled networks, including 'blogs' and 'wikis', whilst a Transition Network Ltd has also been established as a legally constituted body to provide a more formalized structure of support for the movement and its international networking activities.

Finally, the Transition movement deliberately shies away from reifying particular spatial or territorialized boundaries or being overtly prescriptive about which scales work best, which it regards as being a task that is about as easy and as helpful as 'nailing jelly to a wall' (Hopkins and Lipman, 2008: p. 6). Instead it asserts that 'local transition initiatives will identify for themselves the scales that feel most appropriate for them to work at, but [they are encouraged] to work at the scale that feels comfortable and over which they can have an influence' (p. 7). This inherently organic definition of local space, which 'models the ability of natural systems to self-organize' (p. 6), is reflected in the diversity of scope and coverage of transition initiatives in practice. Whilst referred to as 'towns', the communities involved typically cut across and embrace various sub-national spatial scales ranging from villages to council districts, cities and city boroughs.

Nevertheless, its advocates do describe an emerging combination of scales or a multi-layered, nested structure of 'networks in networks' in transition activities. These range from local transition initiatives to regional transition networks and 'hubs', national transition support organizations and networks, as well as temporary groupings of local initiatives to carry out particular projects. Furthermore, there is perceived to be some scope for different scales to fulfil different roles in the evolution and spread of the network and its values. Thus, 'regional hubs' are described as providing a useful mechanism for supporting better links between local initiatives in order 'to help them work synergistically' (p. 12), providing training in key areas such as conflict resolution, organization and other areas 'where local initiatives are too small to provide it effectively' (p. 12). They also provide a useful means for connecting with arms of local and regional government and with coalitions of businesses. Similarly, a range of different 'national networks' has emerged across the different constituent parts of the UK, namely Transition Support Scotland, Transition Ireland Network, Transition England and Transition Support Wales (which is here regarded as a nation), with similar support networks emerging in the USA, New Zealand and across other countries of the world. These are described as playing a key role in embedding the concepts of transition and resilience 'in the language, culture and context of the host nation and/or culture, and also in providing a strategic national overview' (p. 13). They also 'act as ambassadors for the Transition movement at governmental and organizational levels' (p. 13), and provide key infrastructural support relating to systems of coordination and communication, as well as links into key skills and training providers and programmes. In short, regions and nations clearly are viewed as helping to spread the resilience imaginary from the bottom up.

Conclusions

The cultural-political economy approach provides a useful lens for examining competitiveness and its neoliberal foundations and providing a forceful reminder that both of these seemingly irresistible and immovable forces are in fact social constructs that are constantly remoulded and recontextualized to fit different contexts and legitimize different courses of action. As such, whilst they may be currently hegemonic and undeniably robust and variegated, they are not necessarily everlasting and are certainly open to challenge and resistance. Furthermore, the momentum for challenge appears to be gathering force in the context of the triple crunch of economic austerity, climate change and the onset of the era of peak oil which characterizes modernity in the early part of the twenty-first century. Competitiveness and neoliberalism may still be alive and kicking, but they are losing some of their veneer.

The cultural political economic approach also provides some useful insights into the ways and means by which alternative imaginaries might emerge, develop and spread to command the greater rationality required of them to challenge the conventional wisdom. Thus the nature, power and appeal of ideas and interests, and the networks through which these are spread and command support, are thus of critical importance in understanding how and why certain imaginaries are likely to develop and prosper over others. In this regard, this chapter has argued that the imaginary of resilience has particular potential to challenge that of competitiveness, for a number of reasons. It is, potentially, the antithesis of competitiveness inasmuch as it emphasizes activities and policies which are more place based, more cooperative and more diverse. It is also an imaginary around which an array of different interests can coalesce, not least in that it chimes with the emergent concerns of global crisis and the catalytic forces of the triple crunch, as well as with a range of more permissive and progressive grassroots endeavours seeking alternative ways of living and working.

It is also an imaginary which resonates strongly with the emergent rethinking of space and more particularly, how local and regional spaces function. Using the example of the Transition Towns movement, which is pioneering the notion of resilience, this chapter has demonstrated its close connections with the new, broader and more qualitative ontology emerging around the 'region' which conceives the economic imperative – global competitiveness and growth – as only one rationale. In so doing, it has drawn attention to the particular role and function of the 'regional space' in both the battles around competitiveness and the development and spread of alternative imaginaries around resilience.

What emerges from this is strong reinforcement of the view that, although place and context matter, the role and importance of the 'region' specifically cannot be assumed a priori (Hudson, 2007). Regions remain key discursive sites of competition and competitiveness inasmuch as they provide key arenas for legitimizing policy approaches in support of the competitiveness hegemony. Yet at the same time they are increasingly riven by a range of contradictions and paradoxes that make the pursuit of competitiveness in practice a much less clear and deterministic outcome than may previously have been thought. Regions are not necessarily characterized by some unified or notionally shared, collective interest and are not univocal subjects of the policy decisions and mantras developed elsewhere. They are instead characterized by a continuous struggle for coherence between the competing demands of diverse and multiple interests and alliances, with varying values, aspirations and goals. Furthermore, these matters of struggle are rarely uniquely economic in nature or clearly pan-regional in coverage. Instead, they often emanate from localized debates around redistribution and the material conditions of economic growth and uneven development.

As such, the regional scale is uniquely placed – as a sort of 'space in between' – to act as key discursive and strategic, networked sites to scale down the discourse of competitiveness from the 'top down' and scale up alternative practices and imaginaries such as resilience from the 'bottom up'. This helps to explain why regions remain rich sites for institutional and organizational experimentation and restructuring. There is a continuous and ongoing battle to find the best structures for mediating the complex tensions between different interests and communities and between different governments and scales.

Precisely how these tensions are managed and played out will inevitability vary according to the precise relations and balances of power between state, market and civil society in each particular context. Nevertheless, regions have certain capacities to act, albeit at times heavily circumscribed and constrained, and can exercise political choice over their strategic values and direction, which activities they choose to support, and which metrics of development they articulate and promote through their networks.

What is also likely is that action at the local level will only go so far in fostering relevant changes. Nation-states remain dominant actors in pertinent global discussions around the development of new metrics and regulatory regimes for market systems, and play a critical role, through devolution processes, in shaping the powers, resources and capacities available at regional and local levels to enable and empower strong regional action. Securing change on all these different fronts is nothing if not a challenge.

Conclusions

The stated aim of this book was to unpack the concept of regional competitiveness – a concept that has become ubiquitous and dominant in the discourse, policy and practice of regional development. This has inevitably proved to be a broad and far-reaching task which has highlighted both the weaknesses in the theoretical underpinnings of the concept, as well as its limiting implications for policy and practice. In so doing it has shed light on the ways and means by which key discursive constructs such as competitiveness arise and become hegemonic, and has interrogated the very meanings of the 'region' as a space of competitiveness and, subsequently, of alterity. It is, admittedly, only a beginning and there remains much work to be done, particularly in gathering more empirical data on the varieties of competitiveness emerging and on the emerging challenge presented by alternatives. It has nevertheless generated some useful insights into competitiveness, its role, utility and appeal in regional development terms.

One of the principal findings of this study is that competitiveness has become dominant precisely *because* it lacks precision and clarity and is inherently abstract and chaotic. The emergent political economy approaches explored here provide valuable insights for understanding the subsequent malleability of the discourse and thus its utility to both policy makers and the key groups keen to promote the interests it serves. This approach has also highlighted, however, that whilst there is convergence around the *discourse* of competitiveness, *actual* economic and policy practices are much more variegated and diverse. A key endeavour has thus been to try to explore how and why competitiveness is recontextualized in different places, and what this means for its seemingly unending hegemony.

A further key finding is that competitiveness is demonstrably a limiting, growth-first discourse, with the potential to effect and enhance uneven development between places, whilst simultaneously failing to address more fundamental social and ecological matters concerning a place's development. By concentrating on competitiveness, policy makers fail to distinguish between the qualitative aspects of regional development: healthy or unhealthy growth, temporary or sustainable growth. They fail to question what growth is actually needed or what is required actually to improve the quality of life. Thus, policies are developed and perpetuated which tend to promote the wrong kinds of things and which perpetuate invidious forms of place competition.

Powerful technologies of governance, in the form of competitiveness metrics, league tables and benchmarking processes are at work which seek to continually extend the spread of competitiveness and affirm the logic of 'no alternative'. However, these are increasingly being challenged and dismantled from both 'above' and 'below', with a growing momentum evident behind attempts to develop new, more progressive measures of development, progress and capabilities, alongside vibrant localized struggles to develop alternative ways and means of living and working. These appear to be finding particular strength within regions where devolved powers reinforce agonistic political debate and the sense that policy can be 'done differently', and are gathering developing strength from the peculiar conditions and constraints being wrought by the 'triple crunch' of global economic crisis, climate change and the end of plentiful oil supplies.

As has been demonstrated here, however, the scope for the selection and retention of an alternative, more context-specific discourse around regional resilience will critically depend on the relative importance of both the catalytic forces for extra-local change, as well as on the more permissive factors within places. These surround the role of political choice within the constraints of capitalist social relations, and the opportunities for scaling up localized struggles around the competitiveness imperative. As such, there are likely to be multiple, complex and contingent pathways to alternative development strategies.

This in turn reminds us of the need to be reflective about the role of regional space and what this implies for policy. Regions are both relational spaces of flows *and* bounded, territorial spaces. They sit within a more globally integrated, multi-layered and complex institutional architecture. Nevertheless, they remain key sites within which particular kinds of development pathway can be articulated, and where different values and metrics can be debated. They are also increasingly important arenas of political power and choice (Pike et al., 2007; Hudson, 2007). As such, they are uniquely equipped to act as complex mediators or boundary-spanners between the local and the global, between the sub-national and the national, between the micro and the macro (see also Bathelt, 2006). This gives them unique opportunities to resist unhelpful social constructions as well as to develop and spread more progressive ones. This also means that they have important capacities to develop policies which are more sensitive to place and context, which acknowledge and build on the diverse and evolutionary nature of regional development processes, and which build on the politics of the possible, of the alternative, of development paths which meet the broader interests and aspirations of those living and working in regions. They can drive forward the search for new metrics; they can act as key hubs for organizing and supporting localized networks around resilience; they can use the scalar advantages of their buying power to promote re-localization in supply chains and embed new values in public service provision; they can provide a strategic platform for knowledge exchange, collaborative and inclusive working and political debate; and they can also act as ambassadors and advocates of change in their various governance and policy networks.

They are unlikely to be able to do this on their own, however, and it is clear that the capacities of regions to act are often heavily circumscribed. As such, progressive change is likely to demand action from both above and below. Nevertheless the point is that change *is* possible. Competitiveness is a social construct, and as such it is always subject to contest and resistance. New imaginaries around resilience are emerging to challenge its dominance, and these have the potential to flourish in regions that seek to use their strategic and organizational capacities to embed them in their discourses and practices. Alternative regional development pathways are indeed possible. They may ultimately also be necessary and probable.

Notes

2 The political economy of regional competitiveness

1 For further details on the genesis and spread of competitiveness initiatives world-wide see USAID paper 'History of Competitiveness Initiatives', USAID Conference on Building Competitive Advantage in Nations, Budapest 26–28 March 2002, www.cipe.org/pdf/whatsnew/events/budaconf/history.pdf. Accessed 4 May 2009.
2 This publication brings together the reports of the first CAG (1995–96).
3 CAG Mission, http://ec.europa.eu/comm/cdp/cag/mission_en.htm. Accessed 4 May 2009.

4 Performance indicators and rankings

1 See www.imd.ch/wcy. Accessed 4 May 2009.
2 See www.weforum.org/en/initiatives/gcp/index.htm. Accessed May 4 2009.
3 See www.hugginsassociates.com. Accessed 4 May 2009.
4 See www.bertelsmann-stiftung.org/cps/rde/xchg/SID-0A000F0A-701AF6C3/bst_engl/hs.xsl/index.html. Accessed 4 May 2009.
5 See www.stiglitz-sen-fitoussi.fr/en/index.htm. Accessed 4 May 2009.

5 Resisting or restating competitiveness?

1 See: http://wales.gov.uk/icffw/home/?lang = en. Accessed 4 May 2009.

6 The limits to competitiveness

1 The term Red Queen was first coined by Van Valen (1973) referring to Lewis Carroll's *Through the Looking Glass*, where the Red Queen tells Alice 'it takes all the running you can do to keep in the same place'. It has become an established evolutionary hypothesis to describe the constant evolutionary arms race between competing species. Where the effect applies, species keep changing in a never-ending race simply to sustain their current level of fitness such that chaos prevails.

7 Resilient regions

1 See http://transitiontowns.org/TransitionNetwork/TransitionNetwork; There is also a Transition US organization to service the growth of Transition Initiatives: see http://transitionus.org. Accessed 4 May 2009.

Bibliography

ACOA (Atlantic Canada Opportunities Agency) (1996) *Regional Competitiveness: Overview of Recent Research*. New Brunswick, Canada: ACOA.

Adams, N. and Harris, N. (2005) *Best Practice Guidelines for Regional Development Strategies and Spatial Planning in an Enlarged EU*. GRIDS Project, Interreg IIIC. Report by Cardiff University, December.

Allen, J., Massey, D. and Cochrane, A. (1998) *Rethinking the Region*. London: Routledge.

Amin, A. and Thrift, N. (1994) 'Living in the global', in Amin, A. and Thrift, N. (eds) *Globalisation, Institutions and Regional Development in Europe*. Oxford: Oxford University Press, pp. 1–22.

Backlund, A.K. and Sandberg, A. (2002) 'New media industry development: regions, networks and hierarchies – some policy implications', *Regional Studies*, 36, pp. 87–92.

Balanya, B., Doherty, A., Hoedeman, O. and Wesselius, E. (2000) *Europe Inc: Regional and Global Restructuring and the Rise of Corporate Power*. London: Pluto Books.

Barclays, (2002) *Competing with the World: World Best Practice in Regional Economic Development*. London: Barclays Bank plc.

Bathelt, H. (2006) 'Geographies of production: growth regimes in spatial perspective 3 – toward a relational view of economic action and policy', *Progress in Human Geography*, 30 (2), pp. 223–236.

Beacon Hill (2001) *State Competitiveness Report 2001*, Boston: The Beacon Hill Institute for Public Policy Research; Suffolk University.

—— (2007) *State Competitiveness Report 2007*, Boston: The Beacon Hill Institute for Public Policy Research; Suffolk University.

Begg, I. (2002) 'Urban Competitiveness', in Begg, I. (ed.) *Urban Competitiveness: Policies for Dynamic Cities*. Bristol: The Policy Press.

Berger, T. (2008) 'Concepts of national competitiveness', *Journal of International Business and Economy*, 9 (1), pp. 91–112.

Berger, T. and Bristow, G. (2008) 'Benchmarking regional performance: a critical reflection on indices of regional competitiveness', paper for the Regional Studies Association Annual Conference, Prague, May.

BERR (Department for Business, Enterprise and Regulatory Reform) (2008a) *Prosperous Places: Taking forward the Review of Sub-National Economic Development and Regeneration*. BERR: London.

BERR (2008b) *Regional Competitiveness and State of the Regions 2008*. BERR: London (May).

Birch, K. and Mykhnenko, V. (2009) 'Varieties of neoliberalism? Restructuring in large industrially dependent regions across Western and Eastern Europe', *Journal of Economic Geography*, 9, pp. 355–380.

Bishop, K. and Flynn, A. (1999) 'The National Assembly for Wales and the promotion of sustainable development: Implications for collaborative government and governance', *Public Policy and Administration*, 14 (2), pp. 62–76.

—— (2005) 'Sustainable development in Wales: Schemes and structures, debate and delivery', *Contemporary Wales*, 17, pp. 92–112.

Blake, M. and Hanson, S. (2004) 'Rethinking Innovation: Context and Gender', *Environment and Planning A*, 37, pp. 681–701.

Boland, P. (2007) 'Unpacking the theory-policy interface of local economic development: an analysis of Cardiff and Liverpool', *Urban Studies*, 44 (5&6), pp. 1019–1039.

Boschma, R. (2004) 'Competitiveness of regions from an evolutionary perspective', *Regional Studies*, 38, pp. 1001–1014.

Bourdieu, P. (1998) 'The essence of neoliberalism: utopia of endless exploitation'. Translated by J.J. Shapiro. *Le Monde Diplomatique*, Paris (December).

Boyle, D. (2001) *The Tyranny of Numbers: Why Counting Can't Make us Happy.* London: HarperCollins.

Brenner, N. (2000) 'Building "Euro-regions": locational politics and the political geography of neo-liberalism in post-unification Germany', *European Urban and Regional Studies*, 7 (4), pp. 319–345.

—— (2002) 'Decoding the newest "metropolitan regionalism" in the USA: a critical overview', *Cities*, 19, pp. 3–21.

Brenner, N., Jessop, B., Jones, M. and MacLeod, G. (2003) 'Introduction: state space in question', in Brenner, N., Jessop, B., Jones, M. and MacLeod, G. (eds) *State/Space: A Reader.* Oxford: Blackwell, pp. 1–26.

Bristow, G. (2005) 'Everyone's a "winner": problematising the discourse of regional competitiveness', *Journal of Economic Geography*, 5 (3), pp, 285–304.

Bristow, G. and Blewitt, N. (2001) 'The structural funds and additionality in Wales: devolution and multi-level governance', *Environment and Planning A*, 33 (6), pp. 1083–1100.

Bristow, G., Entwistle, T., Hines, F. and Martin, S.J. (2008) 'New spaces for inclusion? Lessons from the "three-thirds" partnerships in Wales', *International Journal of Urban and Regional Research*, 32 (4), pp. 903–921.

Bristow, G., Hartwell, S., Meadows G. and Morgan, K. (2007) *Objective One in Wales: Assessment and Lessons for the Future.* Report for BBC Wales, Cardiff: Cardiff University.

Bristow, G. and Lovering, J. (2006) 'Shaping events, or celebrating the way the wind blows? The role of competitiveness strategy in Cardiff's "ordinary transformation"', in Hooper, A. and Punter, J. (eds) *Capital Cardiff 1965–2020: Regeneration, Competitiveness and the Urban Environment*, Cardiff: University of Wales Press, pp. 311–329.

Bristow, G. and Wells, P. (2005) 'Innovative discourse for sustainable local development: a critical analysis of eco-industrialism', *International Journal of Innovation and Sustainable Development*, 1 (1/2), pp. 168–179.

Bronisz, U., Heijman, W. and Miszczuk, A. (2008) 'Regional competitiveness in Poland: creating an index', in *Jahrbuch für Regionalwissenschaft*, 28 (2), pp. 133–143.

Buck, N., Gordon, I., Harding, A. and Turok, I. (eds) (2005) *Changing Cities: Rethinking Urban Competitiveness, Cohesion and Governance.* Basingstoke: Palgrave Macmillan.

Budd, L. and Hirmis, A.K. (2004) 'Conceptual framework for regional competitiveness', *Regional Studies*, 38 (9), pp. 1015–1028.

Bunzl, J. (2001) *The Simultaneous Policy: An Insider's Guide to Saving Humanity and the Planet*. London: New European Publications.

Burfitt, A., Collinge, C. and MacNeill, S. (2006) 'The discursive construction of regional knowledge economies: a case study of the European Lisbon agenda', paper for the EURODITE meeting in Brussels, 13–15 September 2006.

Burfitt, A. and MacNeill, S. (2008) 'The challenge of pursuing cluster policy in the congested state', *International Journal of Urban and Regional Research*, 32 (2), pp. 492–505.

CAG Consultants (2003) *How Effectively Has the National Assembly for Wales Promoted Sustainable Development?* Report for the Welsh Assembly Government. London: CAG Consultants.

Camagni, R. (2002) 'On the concept of territorial competitiveness: sound or misleading?', *Urban Studies*, 39, pp. 2395–2411.

Cambridge Econometrics, Ecorys-NEI and Martin, R. (2002) *A Study on the Factors of Regional Competitiveness*. Report for the European Commission, Director-General Regional Policy. Brussels: European Commissionn.

Cammack, P. (2006) *The Politics of Global Competitiveness*. Papers in the Politics of Global Competitiveness, No. 1, Manchester: Institute for Global Studies, Manchester Metropolitan University.

Cappellin, R. (1998) 'The transformation of local production systems', in Steiner, M. (ed.) *Clusters and Regional Specialisation: On Geography, Technology and Networks*. London: Pion.

CBI Wales (2002) *Delivering for Wales: CBI Response to the National Assembly's Proposals for Planning*. April. www.cbi.org.uk/wales (accessed 14 September 2008).

Cellini, R. and Soci, A. (2002) 'Pop competitiveness, Banca Nazionale del Lavoro', *Quarterly Review*, 55 (220), pp. 71–101.

Cerny, P.G. (2007) 'Paradoxes of the competition state: the dynamics of globalization', *Government and Opposition*, 32 (2), pp. 251–274.

Chaney, P. and Fevre, R. (2001) 'Inclusive governance and "minority" groups: the role of the third sector in Wales', *Voluntas: International Journal of Voluntary and Nonprofit organisations*, 12 (2), pp. 131–156.

Chien, S.-S. and Gordon, I. (2008) 'Territorial competition in China and the west', *Regional Studies*, 42 (1), pp. 31–49.

Collins, A. (2007) 'Making truly competitive cities – on the appropriate role for local government', *Economic Affairs*, September, pp. 75–80.

Collits, P. (2004) 'Policies for the future of regional Australia', *European Planning Studies*, 12 (1), pp. 85–97.

CEC (Commission of the European Communities) (1993) *White Paper on Growth, Competitiveness and Employment: the Challenges and Ways Forward into the 21st Century*. COM (93) 700 final. Brussels: 5 December 1993.

—— (1999a) *Sixth Periodic Report on the Social and Economic Situation and Development of Regions in the EU*. Brussels: European Commission.

—— (1999b) *Cluster-Building: A Practical Guide*. DG XVI and the Federation of Austrian Industry in Graz, Styria, Austria.

—— (2000) *Lisbon European Council – Presidency Conclusions, 23 and 24 March 2000, Lisbon*. Brussels: European Commission.

—— (2001) *Working for the Regions*. Brussels: European Commission.

—— (2003) *European Competitiveness Report 2003*. Commission Staff Working Paper SEC (2003), 1299. Brussels: 12 November 2003.

—— (2004) *A New Partnership for Cohesion: Convergence, Competitiveness and Cooperation*. Brussels: European Commission.

—— (2005) 'Best practice in regional development', *InfoRegio Panorama*, No. 16, May. Brussels: European Commission.

—— (2006) *Putting Knowledge into Practice: A Broad Based Innovation Strategy for the EU*. Communication from the Commission to the Council, the European Parliament, the European Economic and Social Committee and the Committee of the Regions, COM (2006) 502 final. Brussels: European Commission, 13.09.2006.

—— (2008) *Growing Regions, Growing Europe: Fifth Progress Report on Economic and Social Cohesion*. Communication from the Commission, COM (2008) 371 final: Luxembourg: European Commission, June.

Cooke, P. (2003) 'The regional innovation system in Wales: evolution or eclipse?', in Cooke, P., Heidenreich, M. and Braczyk, H. (eds) *Regional Innovation Systems*. London: Routledge (second edition).

Cooke, P. and Clifton, N. (2005) 'Visionary, precautionary and constrained "varieties of devolution" in the economic governance of the devolved UK territories', *Regional Studies*, 39 (4), pp. 437–451.

Cooke, P.N. and Schienstock, G. (2000) 'Structural competitiveness and learning regions', *Enterprise and Innovation Management Studies*, 1 (1), pp. 265–280.

Council on Competitiveness (2001) *Clusters of Innovation: Regional Foundations of US Competitiveness*, A report by Professor Michael Porter, Harvard Business School for the US Council on Competitiveness: Washington DC.

—— (2005) *Measuring Regional Innovation: A Guidebook for Conducting Regional Innovation Assessments*. Prepared for the US Department of Commerce Economic Development Administration. Washington DC: Council on Competitiveness.

—— (2006) *Regional Innovation: National Prosperity*. Summary report on the Regional Competitiveness Initiative and Proceedings of the 2005 National Summit on Regional Innovation. Prepared for the US Department of Commerce Economic Development Administration. Washington DC: Council on Competitiveness.

Cox, K.R. (2004) 'Globalisation and the politics of local and regional development', *Transactions of the Institute of British Geographers*, 29, pp. 179–194.

Cumbers, A., MacKinnon, D. and McMaster, R. (2003) 'Institutions, power and space: assessing the limits to institutionalism in economic geography', *European Urban and Regional Studies*, 10, pp. 325–342.

Dannestam, T. (2008) 'Rethinking local politics: towards a cultural political economy of entrepreneurial cities', *Space and Polity*, 12 (3), pp. 353–372.

Davies, H.T.O., Nutley, S.M. and Smith, P.C. (eds) (2000) *What Works? Evidence-Based Policy and Practice in Public Services*. Bristol: The Policy Press.

Deas, I. and Giordano, B. (2001) 'Conceptualising and measuring urban competitiveness in major English cities: an exploratory approach', *Environment and Planning A*, 33, pp. 1411–1429.

De Jong, M., Lalenis, K. and Mamadouh, V. (eds) (2002), *The Theory and Practice of Institutional Transplantation: Experiences with the Transfer of Policy Institutions*. Dordrecht: Kluwer Academic Publishers.

De Vol, R.C. (1999) *America's High-Tech Economy: Growth, Development and Risks for Metropolitan Areas*. Santa Monica, CA: Milken Institute.

Doel, M.A. and Hubbard, R.J. (2002) 'Taking world cities literally: marketing the city in a global space of flows', *City*, 6, pp. 351–368.

Doloreux, D. and Parto, S. (2005) 'Regional innovation systems: current discourse and unresolved issues', *Technology in Society*, 27, pp. 133–153.

Donald, B. (2001) 'Economic competitiveness and quality of life in city regions: compatible concepts?', *Canadian Journal of Urban Research*, 10 (2), pp. 259–275.

Donovan, N. and Halpern, D. (2002) *Life Satisfaction: The State of Knowledge and Implications for Government*, Report for Strategy Unit. London: Cabinet Office.

Drahokoupil, J. (2008) 'The rise of the competition state in the Visegrad 4: internationalization of the state as a local project', in B. Van Apeldoorn, J. Drahokoupil and L. Hom (eds) *From Lisbon to Lisbon*. Houndmills: Palgrave Macmillan, pp. 186–207.

DTI (Department of Trade and Industry) (1998) *Our Competitive Future: Building the Knowledge Driven Economy*. White paper. London: TSO.

—— (2001) *Opportunity for All in a World of Change*. London: HMSO.

—— (2003) *Regional Competitiveness and State of the Regions*. London: HMSO.

Dunning, J.H., Bannerman, E. and Lundan, S.M. (1998) *Competitiveness and Industrial Policy in Northern Ireland*, Monograph 5, March. Belfast: Northern Ireland Research Council.

ERT (European Roundtable of Industrialists) (1994) *European Competitiveness: The Way to Growth and Jobs*. Brussels: ERT.

Evans, G. (2003) 'Hard-branding the cultural city – from Prado to Prada', *International Journal of Urban and Regional Research*, 27 (2), pp, 417–440.

Evans, J.L. and Pentecost, J. (1998) 'Economic performance across the UK regions: convergence or divergence?', *Environment and Planning C*, 16 (96), pp. 649–658.

Fairbanks, M. and Lindsay, S. (1997) *Plowing the Sea: Nurturing Sources of Growth in the Developing World*, Boston, MA: Harvard Business School Press.

Fisher, P. (2005) 'Grading places: what do business climate rankings really tell us?' Washington DC: Economic Policy Institute.

Florida, R. (2002) 'Bohemia and economic geography', *Journal of Economic Geography*, 2, pp. 55–71.

—— (2003) *The Rise of the Creative Class: and How It is Transforming Work, Leisure, Community and Everyday Life*. New York: Basic Books.

Fothergill, S. (2005) 'A new regional policy for Britain', *Regional Studies*, 39 (5), pp. 659–667.

Fougner, T. (2006) 'The state, international competitiveness and neoliberal globalisation: is there a future beyond "the competition state"'? *Review of International Studies*, 32, pp. 165–185.

Gardiner, B. (2003) 'Regional competitiveness indicators for Europe – audit, database construction and analysis', paper presented at the Regional Studies Association International Conference, Pisa, 12–15 April.

Gardiner, B., Martin, R. and Tyler, P (2004) 'Competitiveness, productivity and economic growth across the European regions', *Regional Studies*, 38 (9), pp. 1045–1068.

Gassmann, H. (1994) 'From industrial policy to competitiveness policies', *OECD Observer*, 9, pp. 9–10.

Geppart, K. and Stephan, A. (2008) 'Regional disparities in the EU: convergence and agglomeration', *Papers in Regional Science*, 87 (2), pp. 193–217.

Getimis, P. (2003) 'Improving European Union regional policy by learning from the past in view of enlargement', *European Planning Studies*, 11 (1), pp. 77–87.

Gibbs, D.C., Jonas, A.E.G., Reimer, S. and Spooner, D.J. (2001) 'Governance, institutional capacity and partnerships in local economic development: theoretical issues and empirical evidence from the Humber sub-region', *Transactions of the Institute of British Geographers*, 26, pp. 103–119.

Gibson-Graham, J.K. (1996) *The End of Capitalism (as we knew it): A Feminist Critique of Political Economy.* Oxford: Blackwell

Gonzales, S. (2006) 'Scalar narratives in Bilbao: a cultural politics of scale approach to the study of urban policy', *International Journal of Urban and Regional Research*, 30 (4), pp. 836–857.

Goodwin, M., Jones, M. and Jones, R. (2005) 'Devolution, constitutional change and economic development: explaining and understanding the new industrial geographies of the British state', *Regional Studies*, 39, pp. 421–436.

Grant, R.M. (1991) 'Porter's "Competitive advantage of nations": an assessment', *Strategic Management Journal*, 12, pp. 535–548.

Greene, F., Tracey, P. and Cowling, M. (2007) 'Recasting the city into city-regions: Place promotion, competitiveness benchmarking and the quest for urban supremacy', *Growth and Change*, 38 (1), pp. 1–22.

Hadjimichalis, C. (2006) 'Non-economic factors in economic geography and in "new regionalism"; a sympathetic critique', *International Journal of Urban and Regional Research*, 30 (3) pp. 690–701.

Hall, T. and Hubbard, P. (eds) (1998) *The Entrepreneurial City: Geographies of Politics, Regime and Representation.* Chichester: John Wiley.

Harding, A. (2007) 'Taking city-regions seriously? Response to debate on "city-regions": new geographies of governance, democracy and social reproduction', *International Journal of Urban and Regional Research*, 31 (2), pp. 443–458.

Harrison, J. (2006) 'Re-reading the new regionalism: a sympathetic critique', *Space and Polity*, 10 (1), pp. 21–46.

—— (2007) 'From competitive region to competitive city-regions: a new orthodoxy, but same old mistakes', *Journal of Economic Geography*, 7 (3), pp. 311–332.

—— (2008a) 'The region in political economy', *Geography Compass*, 2 (3), pp. 814–830.

—— (2008b) 'Shifting the production of scales: centrally orchestrated regionalism and regionally orchestrated centralism', *International Journal of Urban and Regional Research*, 32 (4), pp. 922–941.

Harvey, D. (1989) *The Condition of Post-Modernity.* Oxford: Blackwell.

—— (2000) *Spaces of Hope.* Edinburgh: University of Edinburgh Press.

Hassink, R. (1993) 'Regional innovation policies compared', *Urban Studies*, 30, pp. 1009–1024.

Haughton, G. and Naylor, R. (2008) 'Reflexive local and regional economic development and international policy transfer', *Local Economy*, 23 (2), 167–178.

Hay, C. and Rosamond, B. (2002) 'Globalisation, European integration and the discursive construction of economic imperatives', *Journal of European Public Policy*, 9, pp. 147–167.

Henderson, H. (1999) *Beyond Globalisation: Shaping a Sustainable Global Economy.* UK: New Economic Foundation and USA: Kumarion Press.

Hendrischke, H. (1999) 'Provinces in competition: region, identity and cultural construction', in Hendrischke, H. and Chongyi, F. (eds) *The Political Economy of China's Provinces: Comparative and Competitive Advantage.* Routledge: London, pp. 1–30.

Hickel, R. (1998) *Standort-Wahn und Euro-Angst.* Hamburg: Rowohlt Verlag.

Hines, C. (2000) *Localization: A Global Manifesto*. London: Earthscan.

Hirschman, A.O. (1977) *The Passions and the Interests: Political Arguments for Capitalism before its Triumph*. Princeton NJ: Princeton University Press.

—— (1986) 'The concept of interest', in Hirschman, A.O. (1986) *Rival Views of Market Society And Other Recent Essays*. Boston MA: Harvard University Press.

HM Treasury (2001) *Productivity in the UK: 3 – The Regional Dimension*. London: HM Treasury.

Holloway, L., Kneafsey, M., Venn, L., Cox, R., Dowler, E. and Tuomainen, H. (2007) 'Possible food economies: a methodological framework for exploring food production – consumption relationships', *Sociologia Ruralis*, 47 (1), pp. 1–19.

Hooper, A. and Punter, J. (eds) (2006) *Capital Cardiff 1965–2020: Regeneration, Competitiveness and the Urban Environment*. Cardiff: University of Wales Press.

Hopkins, R. (2008) *The Transition Handbook: From Oil Dependency to Local Resilience*. Chelsea: Green Books.

Hopkins, R. and Lipman, P. (2008) *The Transition Network Ltd: Who We Are and What We Do*. Version 1.0. Totnes, Devon: Transition Network Ltd.

Hospers, G.J. (2006) 'Silicon somewhere? Assessing the usefulness of best practices in regional policy', *Policy Studies*, 27 (1), pp. 1–15.

House of Commons (2000) *Regional Competitiveness and the Role of the Knowledge Economy*. Research Paper 00/73, 27 July. London: House of Commons

Hubner, D. (2005) 'Regional policy and the Lisbon agenda: challenges and opportunities', speech at the London School of Economics (LSE), London, 3 February.

Hudson, R. (2005) 'Towards sustainable economic practices, flows and spaces: or is the necessary impossible and the impossible necessary?', *Sustainable Development*, 13, pp. 239–252.

—— (2006) 'Regional devolution and regional economic success: myths and illusions about power', *Geografiska Annaler B*, 88, pp. 159–171.

—— (2007) 'Regions and regional uneven development forever? Some reflective comments upon theory and practice', *Regional Studies*, 41 (9), pp. 1149–1160.

—— (2008) 'Cultural political economy meets global production networks: a productive meeting?', *Journal of Economic Geography*, 8 (3), pp. 421–440.

Huggins, R. (2000) *An Index of Competitiveness in the UK: Local, Regional and Global Analysis*. Cardiff: Centre for Advanced Studies, Cardiff University, April.

—— (2003) 'Creating a UK competitiveness index: regional and local benchmarking', *Regional Studies*, 37, pp. 89–96.

Huggins, R. and Day, J. (2005) *UK Competitiveness Index 2005*. Pontypridd: Robert Huggins Associates.

—— (2006) *UK Competitiveness Index 2006*. Pontypridd: Robert Huggins Associates, Work Foundation.

Huggins, R. and Izushi, H. (2008) *UK Competitiveness Index 2008*. Cardiff: Centre for International Competitiveness, UWIC.

IFO (1990) *An Empirical Assessment of Factors Shaping Regional Competitiveness in Problem Regions*. Brussels: Centre for Economic Research (CEC).

IMD (International Institute for Development Management) (1996) *The World Competitiveness Yearbook 1996*. Geneva: IMD.

Institute of Welsh Affairs (IWA) (2001) *World Best Practice in Regional Economic Development*, report for the IWA by ERES Consultants. Cardiff: IWA.

Jackson, T. (2004) *Chasing Progress: Beyond Measuring Economic Growth, The Power of Well-Being*, report for the New Economics Foundation. London: NEF.

—— (2009) *Prosperity Without Growth? The Transition to a Sustainable Economy*, report for the Sustainable Development Commission. London: Sustainable Development Commission (March).

Jacquemin, A. and Pench, L.R. (eds) (1997) *Europe Competing in the Global Economy: Reports of the Competitiveness Advisory Group*. London: Edward Elgar Publishing.

Jarvis, H. (2007) 'Home truths about care-less competitiveness', *International Journal of Urban and Regional Research*, 31 (1), pp. 207–214.

Jayasuriya, K. (1994) 'Singapore: the politics of regional definition', *Pacific Review*, 7 (4), pp. 411–420.

Jeffery, C. and Wincott, D. (2006) 'Devolution in the United Kingdom: statehood and citizenship in transition', *Publius*, 36 (2), pp. 3–18.

Jessop, B. (1994) 'Post-Fordism and the state', in Amin, A. (ed.) *Post-Fordism: A Reader*. Oxford: Blackwell.

—— (2001) 'Institutional (re)turns and the strategic-relational approach', *Environment and Planning A*, 33, pp. 1213–1235.

—— (2004) 'Critical semiotic analysis and cultural political economy', *Critical Discourse Studies*, 1 (2), pp. 159–174.

—— (2005) 'Cultural political economy, the knowledge-based economy and the state', in Barry, A. and Slater, D. (eds) *The Technological Economy*. London: Routledge, pp. 144–166.

—— (2008) 'The cultural political economy of the knowledge-based economy and its implications for higher education', in Jessop, B., Fairclough, N. and Wodak, R. (eds) *Education and the Knowledge-Based Economy in Europe*. Rotterdam: Sense Publishers.

Jessop, B. and Oosterlynck, S. (2008) 'Cultural political economy: on making the cultural turn without falling into soft economic sociology', *Geoforum*, 39 (3), pp. 1155–1169.

Jessop, B. and Sum, N.-L. (2000) 'An entrepreneurial city in action: Hong Kong's emerging strategies in and for (inter)urban competition', *Urban Studies*, 37 (12), pp. 2287–2313.

Jones, C. (2001) 'A level playing field? Sports stadium infrastructure and urban development in the United Kingdom', *Environment and Planning A*, 33, pp. 845–861.

Jones, M. (2008) 'Recovering a sense of political economy', *Political Geography*, 27 (4), pp. 377–399.

Jones, M. and MacLeod, G. (2004) 'Regional spaces, spaces of regionalism: territory, insurgent politics, and the English question', *Transactions of the Institute of British Geographers*, 29, pp. 433–452.

Jones, M., Goodwin, M. and Jones, R. (2005a) 'State modernisation, devolution and economic governance: an introduction and guide to debate', *Regional Studies*, 39. pp. 397–403.

Jones, R., Goodwin, M., Jones, M. and Pett, K. (2005b) '"Filling-in" the state: economic governance and the evolution of devolution in Wales', *Environment and Planning C: Government and Policy*, 23, pp. 337–360.

Keating, M. (1998) *The New Regionalism in Western Europe: Territorial Restructuring and Political Change*. Cheltenham: Edward Elgar.

Keynes, J.M. (1971) *The Collected Writings of John Maynard Keynes*, Vols 1–30. London: Macmillan.

Kitson, M., Martin, R. and Tyler, P. (2004) 'Regional competitiveness: an elusive yet key concept?' *Regional Studies*, 38 (9), pp. 991–999.

Knapp, W. and Schmitt, P. (2003) 'Re-structuring competitive metropolitan regions in North-west Europe: on territory and governance', *European Journal of Spatial Development*, 6 (accessed February 2009: www.nordregio.se/EJSD/refereed6.pdf).

Kok Report (2004) *Facing the Challenge. The Lisbon Strategy for Growth and Employment*, report from the high level group chaired by Wim Kok. Luxembourg: Office for Official Publications of the European Community.

Kresl, P.K. and Singh, B. (1999) 'Competitiveness and the urban economy: twenty four large US metropolitan areas', *Urban Studies*, 36, pp. 1017–1027.

Krueger, R. and Savage, L. (2007) 'City-regions and social reproduction: a "place" for sustainable development', *International Journal of Urban and Regional Research*, 31 (1), pp. 215–223.

Krugman, P. (1994) 'Competitiveness: a dangerous obsession', *Foreign Affairs*, 73 (2), pp. 28–44.

—— (1997a) *Pop Internationalism*. Cambridge MA: MIT Press.

—— (1997b) 'How the economy organises itself in space: A survey of the new economic geography', in Arthur, W.B., Durlauf, S. and Lane, D.A. (eds) *The Economy as an Evolving Complex System 11: Proceedings*, Boulder CO: Westview Press, pp. 223–237.

—— (2003) *Growth on the Periphery: Second Wind for Industrial Regions?* The Allander Series. Scotland: Fraser Allander Institute.

Lagendijk, A. (2007) 'The accident of the region: a strategic relational perspective on the construction of the region's significance', *Regional Studies*, 41 (2), pp.1193–1208.

Lagendijk, A. and Cornford, J. (2000) 'Regional institutions and knowledge – tracking new forms of regional development policy', *Geoforum*, 31, pp. 209–218.

Lall, S. (2001) 'Competitiveness indices and developing countries: an economic evaluation of the Global Competitiveness Report', *World Development*, 29 (9), pp. 1501–1525.

Lambert, R. (2003) *Review of Business-University Collaboration*. Final Report. London: HM Treasury.

Larkin, K. and Cooper, M. (2009) *Into Recession: Vulnerability and Resilience in Leeds, Brighton and Bristol*. London: Centre for Cities (January).

Lawton-Smith, H. Tracey, P. and Clark, G.L. (2003) 'European policy and the regions: a review and analysis of tensions', *European Planning Studies*, 11 (7), pp. 859–873.

Layard, R. (2005) *Happiness: Lessons from a New Science*, London: Allen Lane, Penguin Books.

Leitner, H. and Garner, M. (1993) 'The limits of local initiatives: a reassessment of urban entrepreneurialism for urban development', *Urban Geography*, 14, pp. 57–77.

Leitner, H., Sheppard, E.S., Sziarto, K. and Maringant, A. (2007) 'Contesting urban future: decentering neoliberalism', in Leitner, H., Peck, J. and Sheppard, E.S. (eds) *Contesting Neoliberalism*. New York: Guilford Press, pp. 26–50).

Lever, W.F. and Turok, I. (1999) 'Competitive cities: introduction to the review', *Urban Studies*, 36, pp. 1029–1044.

Leyshon, A., Lee, R. and Williams, C.C. (2003) *Alternative Economic Spaces*. London: SAGE.

Lipietz, A. (1994) 'The national and the regional: their autonomy vis-à-vis the capitalist world crisis', in Palan, R. and Gill, B. (eds) *Transcending the State-Global Divide*, Boulder CO: Lynne Rienner, pp. 23–44.

Lisbon European Council (2000) *Presidency Conclusions*, 23 and 24 March 2000. Brussels: European Commission.

Local Futures Group (2006) *State of the Nation 2006: The Geography of Well-Being in Britain*. London: Local Futures.

Lovering, J. (1999) 'Theory led by policy: the inadequacies of the "new regionalism" (illustrated from the case of Wales)', *International Journal of Urban and Regional Research*, 23, pp. 379–396.

—— (2003) 'MNCs and wannabes: inward investment, discourses of regional development and the regional service class', in Phelps, N.A. and Raines, P. (eds) *The New Competition for Inward Investment: Companies, Institutions and Territorial Development*. Cheltenham: Edward Elgar, pp. 39–60.

McDowell, L. (2004) 'Work, workfare, work/life balance and an ethic of care', *Progress in Human Geography*, 28 (2) pp. 145–163.

MacKinnon, D. Cumbers, A. and Chapman, K. (2002) 'Learning, innovation and regional development: a critical appraisal of recent debates', *Progress in Human Geography*, 26, pp. 293–311.

MacLeod, G. (2001) 'New regionalism reconsidered: globalisation and the remaking of political economic space', *International Journal of Urban and Regional Research*, 25, pp. 804–829.

MacLeod, G. and Jones, M. (2007) 'Territorial, scalar, networked, connected: in what sense a "regional world"?', *Regional Studies*, 41 (9), pp.1177–1191.

Mainwaring, L., Jones, R. and Blackaby, D. (2006) 'Devolution, sustainability and GDP convergence: is the Welsh agenda achievable?', *Regional Studies*, 40 (6), pp. 679–689.

Majone, G. (1989) *Evidence, Argument and Persuasion in the Policy Process*. New Haven: Yale University Press.

Malecki, E. (2002) 'Hard and soft networks for urban competitiveness', *Urban Studies*, 39, pp. 929–945.

—— (2004) 'Jockeying for position: what it means and why it matters to regional development policy when places compete', *Regional Studies*, 38 (9), pp. 1102–1120.

Malmberg, A. and Maskell, P. (2002) 'The elusive concept of localisation economies: towards a knowledge-based theory of spatial clustering', *Environment and Planning A*, 34 (3), pp. 429–449.

Markey, S., Halseth, G. and Manson, D. (2008) 'Closing the implementation gap: a framework for incorporating the context of place in economic development planning', *Local Environment*, 13 (4), pp. 339–351.

Markusen, A. (1994) 'Studying regions by studying firms', *Professional Geographer*, 46, pp. 477–490.

—— (1996) 'Sticky places in slippery space: a typology of industrial districts', *Economic Geography*, 72, pp. 293–313.

—— (2006) 'Economic geography and political economy', in Bagchi-Sen, S. and Lawton-Smith, H. (eds) *Economic Geography: Past, Present, Future*. London: Routledge, pp. 94–102.

—— (2007) (ed.) *Reigning in the Competition for Capital*. Michigan: W.E. Upjohn Institute.

Martin, R. (2005) *Thinking about Regional Competitiveness: Critical Issues*, background 'Think-Piece' paper commissioned by the East Midlands Development Agency, October. Cambridge: University of Cambridge.

Martin, R. and Sunley, P. (2003) 'Deconstructing clusters: chaotic concept or policy panacea?' *Journal of Economic Geography*, 3, pp. 5–35.

Martin, R., Kitson, M. and Tyler, P. (2006) *Regional Competitiveness (Regional Development and Public Policy*. London Routledge.

Maskell, P. and Malmberg, A. (1999) 'Localised learning and industrial competitiveness'. *Cambridge Journal of Economics*, 23, pp. 167–185.

Massey, D. (2000) 'Entanglements of power: reflections', in Sharp, J., Routledge, P., Philo, C. and Paddison, R. (eds) *Entanglements of Power: Geographies of Domination/ Resistance*. London: Routledge, pp. 279–286.

—— (2004) 'The responsibilities of place', *Local Economy*, 19, pp. 97–101.

—— (2005) *For Space*. London: SAGE.

—— (2007a) *World City*. Cambridge: Polity Press.

—— (2007b) 'A new politics of space', *Red Pepper*, August/September, pp. 30–32.

Mayer, H. (2006) 'What is the role of universities in high-tech economic development? The case of Portland, Oregon and Washington DC', *Local Economy*, 21 (3), pp. 292–315.

Mazey, S. and Richardson, J. (1997) 'Policy framing: interest groups and the lead up to the 1996 Inter-Governmental Conference', *West European Politics*, 20 (3), pp. 111–133.

Milliken, J. (1999) 'The study of discourse in international relations: A critique of research and methods', *European Journal of International Relations*, 5 (2), pp. 225–254.

Morgan, K. (1997) 'The learning region: institutions, innovation and regional renewal', *Regional Studies*, 31 (5), pp. 491–503.

—— (2001) 'The new territorial politics: rivalry and justice in post-devolution Britain', *Regional Studies*, 35, pp. 343–348.

—— (2002) 'The English question: regional perspectives on a fractured nation', *Regional Studies*, 36, pp. 797–810.

—— (2004a) 'Sustainable regions: governance, innovation and scale', *European Planning Studies*, 12 (6), 871–889.

—— (2004b) 'The exaggerated death of geography: why proximity still matters for the location of economic activity', *Journal of Economic Geography*, 4 (1), pp. 3–21.

—— (2006) 'Devolution and development: territorial justice and the North-South divide', *Publius: the Journal of Federalism*, 36 (1), pp. 189–206.

—— (2007) 'The polycentric state: new spaces of empowerment and engagement?', *Regional Studies*, 41 (9), pp. 1237–1251.

Morgan, K. and Morley, A. (2006) *Sustainable Public Procurement: Good Practice Case Studies in Wales*. Cardiff: Regeneration Institute, Cardiff University.

Moulaert, F., Martinelli, F., Gonzales, S. and Swyngedouw, E. (2007) 'Introduction: social innovation and governance in European cities – urban development between path dependency and radical innovation', *European Urban and Regional Studies*, 14 (3), pp. 195–209.

Mulderrig, J. (2008) 'Using keywords analysis in CDA: evolving discourses of the knowledge economy in education policy', in Jessop, B., Fairclough, N. and Wodak, R. (eds) *Education and the Knowledge-Based Economy in Europe*. Rotterdam: Sense Publishers.

Mytelka, L.K. and Smith, K. (2002) 'Policy learning and innovation theory: an interactive and co-evolving process', *Research Policy*, 14 (23), pp. 1–13.

National Economics (2008) *State of the Regions 2008–9*, report by National Economics for the Australian Local Government Association. Melbourne: Australian Local Government Association.

NEI (1992) *New location factors for mobile investment in Europe*. Brussels: CEC DG XVI.

New Economics Foundation (2008) *A Green New Deal: Joined-up Policies to Stop the Triple Crunch of the Credit Crisis, Climate Change and High Oil Prices*, report by the New Economics Foundation for the Green New Deal Group. London: Green New Deal Group (July).

—— (2009) *National Accounts of Well-Being*. London: NEF (February).

New South Wales Government (2006) *The State Plan: A New Direction for New South Wales*. Sydney: New South Wales Government.

North, D., Syrett, S. and Etherington, D. (2007) *Devolved Governance and the Economic Problems of Deprived Areas: The Cases of Scotland, Wales and Four English Regions*. York: Joseph Rowntree Foundation.

Nussbaum, M.C. (2000) *Women and Human Development: The Capabilities Approach*. Cambridge: Cambridge University Press.

Ochel, W. and Röhn, O. (2006) *Ranking of countries – the WEF, IMD, Fraser and Heritage indices*. Munich: Centre for Economic Studies, Institute for Economic Research, DICE report 2/2006, pp. 48–60.

OECD (Organisation for Economic Co-operation and Development) (1992) *Technology and the Economy: The Key Relationships*. Paris: OECD, The Technology/Economic Programme.

—— (1996) *Technology, Productivity and Job Creation*. Paris: OECD.

—— (2001) *Innovative Clusters: Drivers of National Innovation Systems*. Pars: OECD.

—— (2005) *Building Competitive Regions: Strategies and Governance*. Paris: OECD.

Ohmae, K. (1995) 'Putting global logic first', *Harvard Business Review*, 73 (1), pp 119–125.

One North East (2006a) *What Works in Regional Economic Development: Learning from International Best Practice*, report by Newcastle University for One North East Development Agency. Newcastle: One North East.

—— (2006b) *Leading the Way: Regional Economic Strategy for North East England 2006–2016*. Newcastle: One North East.

—— (2007) *Globally Competitive Clusters: Creating a Prosperous Future*. Newcastle: One North East.

Paasi, A. (2002) 'Place and region: regional worlds and words', *Progress in Human Geography*, 26 (6), pp. 802–811.

Painter, J. (2008) 'Cartographic anxiety and the search for regionality', *Environment and Planning A*, 40, pp. 342–361.

Palazuelos, M. (2005) 'Clusters: myth or realistic ambition for policy-makers?', *Local Economy*, 20 (2), pp. 131–140.

Pearce, G. and Ayres, S. (2009) 'Governance in the English regions: the role of the Regional Development Agencies', *Urban Studies*, 46 (3), pp. 537–557.

Peck, J. (2003) 'Political economies of scale: fast policy, inter-scalar relations, and neoliberal workfare', *Economic Geography*, 79, pp. 331–360.

Peck, J. and Tickell, A. (1994) 'Jungle law breaks out: neo-liberalism and global–local disorder', *Area*, 26 (4) pp. 317–326.

—— (2007) 'Conceptualizing neoliberalism, thinking Thatcherism', in Leitner, H., Peck, J. and Sheppard, E.S. (eds) *Contesting Neoliberalism*. New York: Guilford Press, pp. 26–50.

—— (2002) 'Neoliberalizing space', *Antipode*, 34; pp. 381–404.

Peet, R. (2001) 'Neoliberalism or democratic development? Review of MacEwan, 1999', *Review of International Political Economy*, 8 (2), pp. 329–343.

—— (2003) *Unholy Trinity: the IMF, World Bank and WTO*. London: Zed Books.

Perrons, D. (2004) 'Understanding social and spatial divisions in the new economy: new media clusters and the digital divide', *Economic Geography*, 80, pp. 45–61.

Petrakos, G., Rodriguez-Pose, A. and Radis, A. (2005) 'Growth, integration and regional disparities in the EU', *Environment and Planning A*, 37 (10), pp. 1837–1855.

Petrella, R. (2000) 'The future of regions: why the competitiveness imperative should not prevail over solidarity, sustainability and democracy', *Geografiska Annaler Series B*, 82 (2), pp. 67–72.

Pike, A., Rodriguez-Pose, A. and Tomaney, J. (2007) 'What kind of local and regional development, for whom?', *Regional Studies*, 41 (9), pp. 1253–1269.

Pinelli, D., Giacometti, R., Lewney, R. and Fingleton, B. (1998) *European Regional Competitiveness Indicators*, Discussion Paper 103. Cambridge: Department of Land Economy, University of Cambridge.

Pollitt, C. (2001) 'Convergence: the useful myth?', *Public Administration*, 79 (4), pp. 933–947.

Pompili, T. (1994) 'Structure and performance of less developed regions in the EC', *Regional Studies*, 28, pp. 679–694.

Porter, M. (1990) *The Competitive Advantage of Nations*. Basingstoke: Macmillan.

—— (1995) 'The competitive advantage of the inner city', *Harvard Business Review*, 74 (May–June), pp. 55–71.

—— (1998) *On Competition*. Cambridge MA: Harvard Business School Press.

—— (2002) *Regional Foundations of Competitiveness: Issues for Wales*. Cambridge MA: Harvard Business School.

—— (2003) 'The economic performance of regions', *Regional Studies*, 37, pp. 549–578.

Porter, M. and Ketels, C.H.M. (2003) *UK Competitiveness: Moving to the Next Stage*, a report for the DTI and ESRC. DTI Economics Paper No. 3, May. London: DTI and ESRC.

Prytherch, D. (2006) 'Narrating the landscapes of entrepreneurial regionalism: rescaling, "new" regionalism and the planned remaking of Valencia, Spain', *Space and Polity*, 10 (3), pp. 203–227.

Purcell, M. (2009) 'Resisting neoliberalisation: communicative planning or counter-hegeemonic movements?', *Planning Theory*, 8 (2), pp. 140–165.

Raco, M. (2008) 'Key worker housing, welfare reform and the new spatial policy in England', *Regional Studies*, 42 (5), pp. 737–751.

Radaelli, C. (2008) 'Europeanization, policy learning and new modes of governance', *Journal of Comparative Policy Analysis*, 10 (3), pp. 239–254.

Reinart, E.S. (1995) 'Competitiveness and its predecessors – a 500 year cross-national perspective', *Structural Change and Economic Dynamics*, 6, pp. 23–42.

Rodaki, N. (2008) 'Recontextualising the competitiveness discourse: branding Rome as a "competitive community"', paper presented to the ESRC Seminar Series on Changing Cultures of Competitiveness: Discourses and Knowledge Brands, Lancaster University, Institute of Advanced Studies and Department of Politics and International Relations, 18 January.

Rodriguez-Pose, A. (1999) 'Innovation prone and innovation averse societies: economic performance in Europe', *Growth and Change*, 30, pp. 75–105.

Rodriguez-Pose, A. and Crescenzi, R. (2008) 'Mountains in a flat world: why proximity still matters for the location of economic activity', *Contemporary Economic Policy*, 1 (3), pp. 371–388.

Rodriguez-Pose, A. and Gill, N. (2005) 'On the "economic dividend" of devolution', *Regional Studies*, 39 (4), pp. 405–420.

Rohr-Zanker, R. (2001) 'How to attract managers and professional to peripheral regions? Recruitment strategies in the Weser-Ems region, Germany', *European Planning Studies*, 9, pp. 47–68.

Romijn, H. and Albu, M. (2002) 'Innovation, networking and proximity: lessons from small high technology firms in the UK', *Regional Studies*, 36 (1), pp. 81–86.

Rosamund, B. (2002) 'Imagining the European economy: "competitiveness" and the social construction of "Europe" as an economic space', *New Political Economy*, 7 (2), pp. 157–177.

Rose, R. (1993) *Lesson-Drawing in Public Policy: A Guide to Learning Across Time and Space*, Chatham: Chatham House Publishers.

Rouvinen (2001) 'Finland on top of the competitiveness game?', *Finnish Economy and Society*, 4/2002, pp. 53–60.

Sanderson, I. (2001) 'Performance management, evaluation and learning in "modern" local government', *Public Administration*, 79 (2), pp. 297–313.

Sanz-Menendez, L. and Cruz-Castro, L. (2005) 'Explaining the science and technology policies of regional governments', *Regional Studies*, 39 (7), pp. 939–954.

Schoenberger, E. (1998) 'Discourse and practice in human geography', *Progress in Human Geography*, 22, pp. 1–14.

Scott, B. and Lodge, G. (eds) (1985) *US Competitiveness and the World Economy*. Boston MA: Harvard Business School Press.

Secretary of State for Wales (1997) *A Voice for Wales*. Cardiff: The Stationery Office.

Sen, A. (1999) *Development as Freedom*. Oxford: Oxford University Press.

Sepic, D. (2004) *Regional Competitiveness: Some Notions*, report for the Russian-European Centre for Economic Policy. Moscow: Russian-European Centre for Economic Policy.

Sheppard, E. (2000) 'Competition in space and between places', in Sheppard, E. and Barnes, T. (eds) *A Companion to Economic Geography*. Oxford: Blackwell.

Simms, A. (2008) *Nine Meals from Anarchy: Oil Dependence, Climate Change and the Transition to Resilience*. London: New Economic Foundation in association with Schumacher North.

Skelcher, C. (2000) 'Changing images of the state: overloaded, hollowed-out, congested', *Public Policy and Administration*, 15 (3), pp. 3–19.

Smith, K. (2001) 'Comparing economic performance in the presence of diversity', *Science and Policy*, 28 (4), pp. 267–276.

Smith, N. (2007) 'Nature as accumulation strategy', in Panitch, L. and Leys, C. (eds) *Coming to Terms with Nature: Socialist Register 2007*. London: Merlin Press, pp. 16–36.

Storper, M. (1995) 'The resurgence of regional economies, ten years later: the region as a nexus of untraded interdependencies', *European Urban and Regional Studies*, 2 (3), pp. 191–221.

—— (1997) *The Regional World*. New York: Guilford Press.

Sum, N.-L. (2004) 'From "integral state" to "integral world economic order": towards a Neo-Gramscian international political economy', Cultural Political Economy Working Paper Series No. 7. Lancaster: Institute for Advanced Studies in Social and Management Sciences, University of Lancaster.

Sum, N.-L. and Jessop, B. (2001) 'The pre- and post-disciplinary perspectives of political economy', *New Political Economy*, 6 (1), pp. 89–101.

Swyngedouw, E.A. (1992) 'The mammon quest. "Glocalisation", interspatial competition and the new monetary order: the construction of new scales', in Dunford, M.

and Kaflakas, G. (eds) *Cities and Regions in the New Europe*. London: Belhaven, pp. 39–67.

Tan, K.G., Kong, Y.T. and Chen, K. (2008) 'Relative competitiveness of 31 mainland China provinces and states of India and ten economies of Association of South East Asian Nations: implications for growth and development', *Competitiveness Review*, 18 (1/2), pp. 87–103.

TaxPayers' Alliance (2008) *The Case for Abolishing the RDAs*, Structure of Government Report No. 3. London: The TaxPayers' Alliance (August).

Taylor, J. and Wren, C. (1997) 'UK regional policy: an evaluation', *Regional Studies*, 31 (9), pp. 835–848.

Tewdr-Jones, M. and Phelps, N.A. (2000) 'Levelling the uneven playing field: inward investment, interregional rivalry and the planning system', *Regional Studies*, 34, p. 429–440.

Thornton, P. (2006) 'North-south divide "starting to close at last"', *The Independent*, 13 November, p. 9.

Thurow, L. (1992) *Head to Head – The Coming Economic Battle Among Japan, Europe and America*. London: Nicholas Brealey.

Trench, A. (ed.) (2007) *Devolution and Power in the United Kingdom*. Manchester: Manchester University Press.

Turner, A. (2001) *Just Capital: The Liberal Economy*. London: Pan.

Turok, I. (2004) 'Cities, regions and competitiveness', *Regional Studies*, 38, pp. 1069–83.

—— (2005) 'Local and national competitiveness: is decentralisation good for the economy?', unpublished working paper. Glasgow: University of Glasgow (June).

UNCTAD (United Nations Conference on Trade and Development) (1995) *Environment, International Competitiveness and Development: Lessons from Empirical Studies*, report by the UNCTAD Secretariat, TD/B/WG.6/10, 12 September. New York: UNCTAD

—— (2004) *Competition, Competitiveness and Development: Lessons from Developing Countries*, report by the UNCTAD Secretariat, UNCTAD/DITC/DITC/CLP/2004/1.

UNICE (1993) *Making Europe more Competitive: Towards World Class Performance, an Interim Report*. Brussels: UNICE.

—— (1994) *Making Europe Competitive: Towards World Class Performance. The UNICE Competitiveness Report*. Brussels: UNICE.

—— (1997) *Benchmarking Europe's Competitiveness*. Brussels: UNICE.

Unwin, C. (2006) *Urban Myth: Why Cities Don't Compete*. Discussion Paper No. 5. Centre for Cities, Institute of Public Policy Research (IPPR), February. London: Centre for Cities

Van Valen, L. (1973) 'A new evolutionary law', *Evolutionary Theory*, 1 (1), pp. 1–30.

Varney, D. (2008) *Review of the Competitiveness of Northern Ireland*, report for HM Treasury. London: HM Treasury.

Walker, D. (2002) *In Praise of Capitalism: A Critique of the New Localism*. London: Catalyst.

Ward, K. and Jonas, A.E.G. (2004) 'Competitive city-regionalism as a politics of space: a critical reinterpretation of the new regionalism', *Environment and Planning A*, 36, pp. 2119–2139.

Weaver, P. and Hudson, R. (1995) *Economic Restructuring and Public Expenditure for Sustainable Development: an Eco-Keynesian Model*. London: Working Futures, Friends of the Earth.

Webb, D. and Collis, C. (2000) 'Regional development agencies and the "new regionalism" in England', *Regional Studies*, 34, pp. 857–873.

Welsh Assembly Government (WAG) (2002) *A Winning Wales: the National Economic Development Strategy of the Welsh Assembly Government*. Cardiff: Welsh Assembly Government.s

—— (2003) *Wales for Innovation: The Welsh Assembly Government's Action Plan for Innovation*. Cardiff: Welsh Assembly Government.

—— (2004) *People, Places and Futures*: *The Wales Spatial Plan*. Cardiff: Welsh Assembly Government.

—— (2006) *WAVE: Wales A Vibrant Economy*. The Welsh Assembly Government's Strategic Framework for Economic Development. Cardiff: Welsh Assembly Government.

Williams, R. (2005) *Culture and Capitalism*. London: Verso.

Yeatman, A. (ed.) (1998) *Activism and the Policy Process*. Sydney, Australia: Allen and Unwin.

Zukin, S. (1995) *The Cultures of Cities*. Cambridge MA: Blackwell.

Index of authors

Index of subjects